Advanced Statistical Analytics for Health Data Science with SAS and R

In recent years, there has been a growing emphasis on making statistical methods and analytics accessible to health data science researchers and students. Following the first book on "Statistical Analytics for Health Data Science with SAS and R" (2023, www.routledge.com/9781032325620), this book serves as a comprehensive reference for health data scientists, bridging fundamental statistical principles with advanced analytical techniques. By providing clear explanations of statistical theory and its application to real-world health data, we aim to equip researchers with the necessary tools to navigate the evolving landscape of health data science.

Designed for advanced-level data scientists, this book covers a wide range of statistical methodologies, including models for longitudinal data with time-dependent covariates, multi-membership mixed-effects models, statistical modeling of survival data, Bayesian statistics, joint modeling of longitudinal and survival data, nonlinear regression, statistical meta-analysis, spatial statistics, structural equation modeling, latent growth curve modeling, causal inference, and propensity score analysis.

A key feature of this book is its emphasis on real-world applications. We integrate publicly available health datasets and provide case studies from a variety of health applications. These practical examples demonstrate how statistical methods can be applied to solve critical problems in health science.

To support hands-on learning, we offer implementation guidance using SAS and R, ensuring that readers can replicate analyses and apply statistical techniques to their own research. Step-by-step computational examples facilitate reproducibility and deeper exploration of statistical models. By combining theoretical foundations with practical applications, this book empowers health data scientists to develop robust statistical solutions for complex health challenges. Whether working in academia, industry, or public health, readers will gain the expertise to advance data-driven decision-making and contribute to evidence-based health research.

Chapman & Hall/CRC Biostatistics Series
Series Editors: Mark Chang, *Boston University, USA*

Value of Information for Healthcare Decision Making
Edited by Anna Heath, Natalia Kunst, and Christopher Jackson

Probability Modeling and Statistical Inference in Cancer Screening
Dongfeng Wu

Development of Gene Therapies
Strategic, Scientific, Regulatory, and Access Considerations
Edited by Avery McIntosh and Oleksandr Sverdlov

Bayesian Precision Medicine
Peter F. Thall

Statistical Methods for Dynamic Disease Screening and Spatio-Temporal Disease Surveillance
Peihua Qiu

Causal Inference in Pharmaceutical Statistics
Yixin Fang

Applied Microbiome Statistics
Correlation, Association, Interaction and Composition
Yinglin Xia and Jun Sun

Association Models in Epidemiology
Study Design, Modeling Strategies, and Analytic Methods
Hongjie Liu

Likelihood Methods in Survival Analysis
with R Examples
Jun Ma, Malcolm Hudson, and Annabel Webb

Biostatistics for Bioassay
Ann Yellowlees and Matthew Stephenson

Power and Sample Size in R
Catherine M. Crespi

R for Health Technology Assessment
Gianluca Baio, Howard Thom, and Petros Pechlivanoglou

Advanced Statistical Analytics for Health Data Science with SAS and R
Ding-Geng Chen and Jeffrey R. Wilson

For more information about this series, please visit: www.routledge.com/Chapman--Hall-CRC-Biostatistics-Series/book-series/CHBIOSTATIS

Advanced Statistical Analytics for Health Data Science with SAS and R

Ding-Geng Chen and Jeffrey R. Wilson

CRC Press
Taylor & Francis Group
A CHAPMAN & HALL BOOK

First edition published 2026
by CRC Press
2385 NW Executive Center Drive, Suite 320, Boca Raton FL 33431

and by CRC Press
4 Park Square, Milton Park, Abingdon, Oxon, OX14 4RN

CRC Press is an imprint of Taylor & Francis Group, LLC

ISBN: 9781032978499 (hbk)
ISBN: 9781032979625 (pbk)
ISBN: 9781003596370 (ebk)

DOI: 10.1201/9781003596370

Typeset in Palatino
by Newgen Publishing UK

Contents

Preface ... ix
Acknowledgments ... xiii
About the Authors ... xv

12. Marginal Models for Binary Longitudinal Outcomes with Time-Dependent Covariates ... 1
12.1 Research Interest/Question ... 1
12.2 GEE Estimation for Time-Dependent Covariates ... 1
12.3 GMM Estimation for Time-Dependent Covariates ... 3
12.4 Extended Classification of Systematic Component Models for Time-Dependent Covariates ... 5
12.5 Joint Modeling for Correlated Binary Responses ... 7
12.6 GMM Estimation for Time-Dependent Covariate Models ... 13
12.7 Simulation Study ... 16
12.8 Numerical Example ... 17
12.9 Discussions and Comments ... 21

13. Multiple Membership Mixed-Effects Models ... 25
13.1 Research Interest/Question ... 25
13.2 Hospital Doctor Patient Data with Multiple Membership ... 26
13.3 Two-Level Linear Mixed-Effects Models for Multiple Membership Data ... 28
13.4 Data Analysis Using SAS ... 29
13.5 Data Analysis Using R ... 34
13.6 Discussions and Comments ... 37
13.7 Exercises ... 38

14. Statistical Modeling of Survival Data ... 41
14.1 Research Interest/Question ... 41
14.2 HIV/AIDS Data ... 42
14.3 Principles of Survival Data Analysis ... 42
14.4 Functions of the Distribution ... 44
14.5 Survival Models ... 45
14.6 Data Analysis: Time-to-Death Study ... 51
14.7 Exercises ... 57

15. Statistical Modeling with Bayesian Paradigm ... 59
15.1 Research Interest/Question ... 59
15.2 Public Health NHANES Data ... 59
15.3 Bayes Theorem ... 60

15.4 Bayesian Modeling ... 60
15.5 Bayesian Data Analysis: Continuous Outcome ... 64
15.6 Bayesian Data Analysis: Binary Outcome ... 74
15.7 Summary and Comments ... 80
15.8 Bayesian Analysis: Advantages and Disadvantages ... 80
15.9 Exercises ... 81

16. Jointly Modeling to Analyze Longitudinal and Survival Data with Bayesian Approach ... 83
16.1 Research Interest/Question ... 83
16.2 HIV/AIDS Data and Preliminary Data Analysis ... 84
16.3 Statistical Joint Modeling ... 85
16.4 Data Analysis Results ... 89
16.5 Conclusions and Recommendations ... 93
16.6 Exercises ... 94

17. Nonlinear Regression ... 97
17.1 Research Interest/Questions ... 97
17.2 Chlorine Data ... 97
17.3 Nonlinear Regression Models ... 98
17.4 Fitting Nonlinear Models ... 99
17.5 Data Analysis with SAS and R ... 101
17.6 Results and Discussions ... 106
17.7 Exercises ... 107

18. Statistical Meta-analysis ... 108
18.1 Research Interest/Questions ... 108
18.2 Metadata on Bacillus Calmette–Guerin Vaccine Clinical Trials ... 109
18.3 Meta-analysis Models: Fixed Effects and Random Effects ... 110
18.4 Meta-regression ... 114
18.5 Meta-analysis Using SAS ... 115
18.6 Meta-analysis Using R ... 122
18.7 Summary and Discussions ... 131
18.8 Exercises ... 132

19. Spatial Statistical Analysis ... 134
19.1 Research Interest/Questions ... 134
19.2 New York Leukemia Data and Meuse River Data ... 135
19.3 Models in Spatial Statistical Analysis ... 136
19.4 Spatial Analysis of Leukemia Data and Meuse River Data in SAS/R ... 139
19.5 Discussions and Conclusions ... 155
19.6 Exercises ... 157

20. Structural Equation Modeling ... 158
20.1 Research Interest/Questions ... 158
20.2 Asthma Data ... 158
20.3 Structural Equation Modeling ... 160
20.4 Confirmatory Factor Analysis: Validity of Latent Constructs ... 162
20.5 SEM Versus Regression Analysis ... 163
20.6 Analysis of NHANES Data ... 164
20.7 Discussions ... 178
20.8 Exercises ... 178

21. Longitudinal Data Analysis and Latent Growth Curve Modeling ... 180
21.1 Research Interest/Questions ... 180
21.2 Breast Cancer Data with Measure Patterns of Change in Personality Assessment ... 181
21.3 Longitudinal Data Analysis for Mood and Social Adjustment ... 183
21.4 Joint Latent Class Growth Curve Modeling ... 191
21.5 Conclusions and Discussions ... 202
21.6 Exercises ... 203

22. Latent Growth Mixture Joint Modeling in Intervention Research ... 205
22.1 Research Interest/Question ... 205
22.2 Motivation for a Latent Growth Mixture Joint Modeling Approach ... 206
22.3 Real Data ... 208
22.4 A Step-by-Step Demonstration Using R for Data Analyses ... 209
22.5 Discussions and Conclusions ... 219
22.6 Exercises ... 221

23. Causal Inference and Propensity Score Analysis ... 223
23.1 Research Interest/Questions ... 223
23.2 German Breast Cancer Data ... 224
23.3 Propensity Score Analysis ... 224
23.4 Propensity Score Analysis Using SAS and R ... 225
23.5 Discussions ... 238
23.6 Exercises ... 238

Index ... 241

Preface

The first book, *Statistical Analytics for Health Data Science with SAS and R* (2023, http://www.routledge.com/9781032325620), was well-received and has been widely utilized in workshops and courses at Arizona State University. Based on requests from students and workshop attendees for the inclusion of more advanced topics, we are pleased to present this second book, *Advanced Statistical Analytics for Health Data Science with SAS and R*. This new book addresses the demand for advanced statistical methods and models, offering an expanded suite of tools and techniques designed for health data sciences but broadly applicable across disciplines.

The objective of this volume is to equip readers with advanced analytical tools to address complex data challenges. While the primary focus is on applications in health and health-related data, the models and methods presented are applicable to a wide range of data domains. This book uses R and SAS, two widely adopted statistical software platforms, with accompanying datasets and computational programs provided to enhance the learning experience. The target audience includes applied data scientists, public health professionals, academic researchers, graduate students in statistics and biostatistics, and R&D practitioners in industry and governmental agencies.

The **first book** introduced fundamental statistical methods in 11 chapters, with the initial eight chapters covering independent data and covariates, and the latter three chapters addressing models for correlated data. This **second book** builds upon that foundation, delving into advanced topics and methodologies that expand the scope of health data analytics.

Overview of Chapters

- **Chapter 12: Marginal Models for Binary Longitudinal Outcomes with Time-Dependent Covariates**

  Introduces a generalized method of moments (GMM) approach for analyzing binary longitudinal outcomes with time-dependent covariates, using partitioned coefficients to provide detailed insights into covariate effects over time.

- **Chapter 13: Multiple Membership Mixed-Effects Models**

  Expands the single-membership mixed-effects models discussed in the first book to address the more complex multimembership structures often encountered in healthcare settings, where patients receive care from multiple providers.

- **Chapter 14: Statistical Modeling of Survival Data**

  Discusses survival analysis methods, including Kaplan–Meier estimation, life tables, and Cox proportional hazards regression, with a focus on nonparametric approaches.

- **Chapter 15: Statistical Modeling with Bayesian Paradigm**

  Introduces Bayesian methods as an alternative to the frequentist approach, providing a framework for incorporating prior information and updating inferences as new data become available.

- **Chapter 16: Joint Modeling to Analyze Longitudinal and Survival Data with Bayesian Approach**

  Examines joint modeling techniques for simultaneous analysis of longitudinal and survival data in clinical research, providing more precise treatment effect estimates.

- **Chapter 17: Nonlinear Regression**

  Explores nonlinear regression models to address scenarios where relationships cannot be adequately captured by linear models.

- **Chapter 18: Statistical Meta-analysis**

  Introduces meta-analysis methods, including fixed-effects and random-effects models, and discusses metaregression to account for between-study heterogeneity in systematic reviews.

- **Chapter 19: Spatial Statistical Analysis**

  Provides an introduction to spatial statistics with applications to real-world datasets, such as the New York leukemia and Meuse River data.

- **Chapter 20: Structural Equation Modeling**

  Discusses SEM as a comprehensive framework combining factor analysis and regression to model relationships between latent variables and observed data.

- **Chapter 21: Longitudinal Data Analysis and Latent Growth Curve Modeling**

  Explores latent growth curve modeling, with applications to longitudinal data, including a case study of mood and social adjustment in women recovering from breast cancer surgery.

- **Chapter 22: Latent Growth Mixture Joint Modeling in Intervention Research**

  Examines latent growth mixture modeling, enabling the analysis of heterogeneous trajectories and addressing challenges such as outcome-dependent missing data.
- **Chapter 23: Causal Inference and Propensity Score Analysis**
  Introduces causal inference frameworks and propensity score methods to adjust for confounding in estimating treatment effects.

Acknowledgments

We extend our deepest gratitude to the graduate students at Arizona State University, whose questions and insights have enriched the content of this book. We also thank the faculty and staff of the Department of Economics, the W. P. Carey School of Business, and the College of Health Solutions for their support throughout this process.

This book is dedicated to our students—past and present—whose intellectual curiosity has inspired and motivated us to expand upon the foundational material of **these two books**. To all who contributed to the completion of this work, we express our sincere appreciation.

Ding-Geng Chen and Jeffrey R. Wilson
Arizona State University, Phoenix, AZ, USA

About the Authors

Ding-Geng Chen (aka Din Chen) is an elected fellow of the American Statistical Association (ASA) and an elected member of the International Statistical Institute (ISI). Currently, he is the executive director and professor in biostatistics at the College of Health Solutions, Arizona State University. He is also an elected member of the Academy of Science of South Africa (MASSAf), an extraordinary professor and the South Africa Research Chair (SARChI) in Biostatistics at the University of Pretoria, an honorary professor at the University of KwaZulu-Natal, South Africa. He was the Wallace H. Kuralt distinguished professor in Biostatistics at the University of North Carolina at Chapel Hill, a professor in biostatistics at the University of Rochester Medical School, and the Karl E. Peace Endowed Eminent Scholar Chair in Biostatistics at Georgia Southern University. He is a senior biostatistics consultant for biopharmaceuticals and government agencies with extensive expertise in biostatistics, clinical trials, and public health statistics. He has more than 250 referred professional publications and coauthored and coedited 42 books on clinical trial methodology, meta-analysis, data science, causal inference, and public health research.

Jeffrey R. Wilson is Professor of Statistics and Biostatistics and the Associate Dean of Research and Inclusive Excellence. He is the recipient of several prestigious honors, including the 2024 Dr. Martin Luther King Jr. Faculty Service-Leadership Award, the 2022 APHA Statistical Section Lifetime Service Award, and the 2021 Faculty Achievement Award on Founders' Day. Previously, he served as Director of the School of Health Management and Policy and Chair of the Editorial Board for the *American Journal of Public Health*. Dr. Wilson has held significant leadership roles and has published over 100 articles in leading journals. His research has been supported by major funding agencies, including NIH, NSF, and USDA. As a sought-after consultant for pharmaceutical companies, hospitals, and legal experts, he has also taught advanced statistics and biostatistics courses at Mayo Clinic, Barrett Honors College, and internationally at Zayed University and the University of Pretoria. Dr. Wilson holds a PhD in statistics from Iowa State University and a BA in Mathematics from the University of the West Indies.

12

Marginal Models for Binary Longitudinal Outcomes with Time-Dependent Covariates

12.1 Research Interest/Question

In this chapter, we present and discuss a simultaneous generalized method of moments (GMM) model for analyzing multiple outcomes in longitudinal studies, with partitioned coefficients to account for time-dependent covariates. This innovative approach leverages valid moment conditions to ensure efficient parameter estimation, while the partitioned coefficients offer detailed insights into the effects of each covariate on the outcomes over time.

By concatenating the valid moment conditions, we extend the dimension of the objective function to incorporate the correlation between multiple outcomes, thereby improving model accuracy. A key advantage of this marginal modeling approach is that it avoids reliance on additional distributional assumptions typically required in joint modeling frameworks.

To demonstrate the utility of this method, we apply it to simultaneously examine the risk factors associated with smoking, social alcohol consumption, and obesity among adolescents in the United States. Additionally, we provide comparisons with separately fitted models to highlight the impact of accounting for correlations between the outcome variables.

12.2 GEE Estimation for Time-Dependent Covariates

Longitudinal data are widely used across various disciplines to study how individual responses change over time. This repeated-measures framework inherently introduces correlation, not only among repeated measurements on the same individual but also between time-dependent covariates and outcomes across different time points. Additionally, researchers are often

DOI: 10.1201/9781003596370-1

interested in examining multiple outcomes simultaneously, which further introduces correlations between these outcomes. For instance, the National Longitudinal Study of Adolescent to Adult Health (Add Health) tracks students beginning in 1994–1995, with follow-ups at three subsequent time points (Harris and Udry 2022). In this study, researchers may wish to simultaneously investigate correlated outcomes, such as smoking and alcohol use. Time-dependent covariates, such as depression level and physical activity, could have both immediate and carry-over effects on these outcomes. An appropriate modeling framework must address these layers of correlation to produce valid inferences.

In this context, we refer to the separate outcome variables as outcomes, while the repeated measurements of each outcome are referred to as responses. In this chapter, we present a simultaneous *GMM* model that accounts for multiple outcomes as well as the time dependency inherent in longitudinal data.

12.2.1 Marginal Models and Time-Dependent Covariates

Marginal models for single responses with time-independent covariates have been extensively studied. However, models addressing time-dependent covariates remain an active area of research. Generalized estimating equations (GEEs) (Zeger and Liang 1986), a popular marginal model for correlated data, use a working correlation structure to capture associations between repeated measurements. Yet, studies by Hu et al. (1998) and Pepe and Anderson (1994) demonstrated that GEE models with arbitrary working correlation structures may produce inconsistent estimates when time-dependent covariates are present. They recommended using an independent working correlation structure to mitigate this issue.

Lai and Small (2007) introduced GMM as an alternative to GEE for marginal models with time-dependent covariates. Their approach classified time-dependent covariates into three types and used appropriate estimating equations that improved efficiency compared to GEE with an independent correlation structure. Subsequent advancements by Zhou et al. (2014) and Lalonde et al. (2014) further refined the GMM framework by modifying moment conditions and allowing flexibility in handling covariate characteristics.

Despite these advancements, traditional GMM models assume that the effects of covariates on responses remain constant over time. This assumption is often unrealistic in practice. To address this limitation, Irimata et al. (2019) proposed a partitioned GMM model, which allows the effect of each time-dependent covariate to vary across time. Their approach extended the functional model discussed by Muller and Stadtmuller (2005), offering more detailed insights into covariate effects over time. However, their model was limited to single outcomes and did not account for correlations between multiple outcomes.

12.2.2 Multivariate Longitudinal Models

In many fields, multivariate longitudinal data are common. Separate models for each response can be fit, but such approaches ignore interdependencies between multiple outcomes, potentially leading to inefficient parameter estimates (Berridge and Crouchley 2011; Fitzmaurice et al. 2004). This oversight can result in biased conclusions regarding covariate effects.

Multivariate longitudinal modeling (Bandyopadhyay et al. 2011; Fieuws and Verbeke 2004; Fieuws et al. 2007) addresses this limitation by using multivariate distributions to account for correlations between multiple outcomes. These methods improve efficiency and reduce bias (Gueorguieva 2001; McCulloch 2008). However, their reliance on correctly specified distributions poses challenges, as misspecification may invalidate inferences (Wu et al. 2012).

Joint modeling approaches have also been proposed to address interdependencies between multiple outcomes. These models often link outcomes through random effects (Liu et al. 2010; Ghebremichael 2015) or employ semiparametric frameworks (Song et al. 2002). Joint latent class models (Proust-Lima 2014) and multiple-equation GMM methods in econometrics (Hayashi 2000) have also been used to address correlations between outcomes. However, these approaches often fail to accommodate time-dependent covariates or require additional distributional assumptions.

12.2.3 Simultaneous GMM Approach

To address these gaps, we present a joint modeling approach that simultaneously estimates regression equations for multiple outcomes while accounting for time-varying effects of covariates. Unlike traditional multivariate approaches, our method:

1. does not require distributional assumptions, ensuring broader applicability;
2. incorporates time-varying effects by partitioning regression parameters, capturing the dynamic impact of covariates;
3. simultaneously addresses the interdependence of multiple outcomes and correlations across time points.

12.3 GMM Estimation for Time-Dependent Covariates

The GMM estimator is widely used in econometrics (Hansen 1982) and has gained significant traction in statistical modeling due to its flexibility and robustness. Researchers have demonstrated the efficacy of GMM in handling time-dependent covariates by using a single regression parameter for each covariate (Lai and Small 2007; Lalonde et al. 2014; Chen and Westgate 2017). Building on this work, Irimata et al. (2019) introduced a partitioned GMM model that allows multiple regression coefficients per covariate, providing deeper insight into how the effects of time-dependent covariates vary over time.

12.3.1 The GMM Framework

Consider a study involving N subjects with repeated observations over T time periods and J covariates. Let y_{it} denote the outcome measured for the ith subject at the tth time period, with marginal density given the time-varying vector of covariates $\mathbf{x}_{it.}$, following a generalized linear model. Observations y_{is} and y_{kt} are assumed to be independent when $i \neq k$ but correlated when $i = k$.

The conditional expectation of y_{it}, given the vector of covariates $\mathbf{x}_{it.}$ and regression parameters β, is denoted by $\mu_{it}(\boldsymbol{\beta})$. Following Lai and Small (2007), the GMM estimation relies on the following moment condition:

$$E\left[\frac{\partial \mu_{is}(\boldsymbol{\beta})}{\partial \beta_j}\{y_{it} - \mu_{it}(\boldsymbol{\beta})\}\right] = 0, \tag{12.1}$$

for appropriately selected s, t, and j.

12.3.2 Single-Parameter GMM Models

GMM models using a single regression parameter for each covariate have become increasingly common (Lai and Small 2007; Zhou et al. 2014; Chen and Westgate 2017; Qu et al. 2000). For clarity, this section primarily focuses on the framework established by Lalonde et al. (2014), although the methods discussed can be extended to accommodate alternative approaches.

12.3.3 Covariate Classification

Lai and Small (2007) introduced a GMM marginal model for time-dependent covariates by classifying covariates into three types—Type I, Type II, and Type III—based on the validity of the moment conditions described in Equation (12.1):

- Type I covariates: These satisfy Equation (12.1) for all s and t, making all moment conditions valid.
- Type II covariates: These satisfy Equation (12.1) only when $s \geq t$, but not for $s < t$. An example includes covariates that are not independent of future responses but are unaffected by feedback from the response.

- Type III covariates: These do not satisfy Equation (12.1) for any $s > t$, often due to feedback processes from the outcome to the covariate.

This classification provided a structured approach to selecting appropriate moment conditions for estimating regression parameters.

12.3.4 Hypothesis Testing for Moment Conditions

Lalonde et al. (2014) extended the GMM model by introducing hypothesis testing to validate each moment condition separately. This approach eliminates the need for covariate classification by testing the validity of each moment condition directly. The validity of Equation (12.1) is equivalent to the correlation between the residual e at time s and the covariate x at time t being zero, i.e., $\rho_{x_t, e_s} = 0$. They assumed that all moment conditions for $s = t$ were valid and tested the remaining $T(T - 1)$ moment conditions under the null hypothesis: H_0: $\rho_{x_t, e_s} = 0$. This hypothesis was evaluated using a Z-statistic based on the pairwise correlations between residuals and covariates at each time point. Moment conditions deemed valid under the null hypothesis were incorporated into the regression model estimation.

12.3.5 Advantages and Limitations

This hypothesis-testing approach offers a significant advantage by utilizing only valid moment conditions, which improves the robustness of the model without requiring prior assumptions about covariate classification. However, it is limited by its reliance on the pairwise testing of correlations, which may not fully capture the complexity of feedback processes in certain settings. The single-parameter GMM framework provides a strong foundation for analyzing time-dependent covariates, but its inability to accommodate varying effects over time limits its applicability in complex settings. The next section introduces a partitioned GMM model, which addresses this limitation by allowing time-varying effects for each covariate while maintaining the desirable properties of the GMM estimator.

12.4 Extended Classification of Systematic Component Models for Time-Dependent Covariates

Single-parameter-per-covariate GMM models are effective for analyzing time-dependent covariates but impose the restrictive assumption that each covariate's effect remains constant across all time periods. In practice, this may not adequately capture the dynamic relationships between covariates and responses over time. To address this limitation, Irimata et al. (2019) proposed a partitioned GMM model that employs multiple parameters for each time-dependent covariate, allowing the relationships between covariates and responses to vary over time.

12.4.1 Partitioned GMM Model Framework

This approach uses a partitioning method based on the time lag between the covariate and the response. Moment conditions are grouped according to this partitioning, and a lower diagonal design matrix is constructed for each covariate $\mathbf{X}_{ij}$:

$$\mathbf{X}_{ij} = \begin{bmatrix} 1 & X_{ij1} & 0 & \cdots & 0 \\ 1 & X_{ij2} & X_{ij1} & \cdots & 0 \\ \vdots & \vdots & \vdots & \cdots & \vdots \\ 1 & X_{ijT} & X_{ij(T-1)} & \cdots & X_{ij1} \end{bmatrix} = \begin{bmatrix} \mathbf{1} & X_{ij}^{[0]} & X_{ij}^{[1]} & \cdots & X_{ij}^{[T-1]} \end{bmatrix}$$

This matrix is incorporated into the marginal model:

$$g(\mu_{it}) = \beta_0 + \beta_j^{tt} X_{ij}^{[0]} + \beta_j^{[1]} X_{ij}^{[1]} + \beta_j^{[2]} X_{ij}^{[2]} \ldots + \beta_j^{[T-1]} X_{ij}^{[T-1]} \tag{12.2}$$

where $g(\cdot)$ is the appropriate link function for subject i.

- β_j^{tt} represents the cross-sectional effect of the jth covariate on the outcome when observed within the same time period.
- $\beta_j^{[1]}, \ldots \beta_j^{[T-1]}$ represent the effects of the jth covariate on the outcome observed across a 1 to $T - 1$ time-period lag, respectively. For example, $\beta_j^{[1]}$ captures the relationship between $\mathbf{X}_{ij}$ and the outcome across a one-time-period lag.

This flexible parameterization allows the model to account for the evolving relationships between covariates and responses over time, which is especially valuable in studies where time-lagged effects are of primary interest.

12.4.2 Identification of Valid Moment Conditions

The partitioned GMM model identifies valid moment conditions using either:

1. The Type II covariate framework introduced by Lai and Small (2007), which identifies covariates that satisfy moment conditions for $s \geq t$, or

2. The hypothesis testing approach introduced by Lalonde et al. (2014), which tests the validity of each moment condition through hypothesis testing.

Simulation studies by Irimata et al. (2019) indicated that the Lalonde method tends to produce less biased estimates and better parameter coverage compared to the Lai and Small framework, making it a more robust approach in many settings.

12.4.3 Model Applicability and Extensions

The partitioned GMM model is best suited to datasets with a relatively small number of time periods compared to the number of subjects. This restriction aligns with limitations found in GEE models (Stoner et al. 2010). The framework is particularly advantageous in settings where the dynamic nature of covariate effects across time must be captured without oversimplification.

Building on this foundation, the framework can be extended to account for multiple simultaneous responses of interest by employing a joint model fitting procedure. This extension enables the analysis of correlated outcomes, further enhancing the utility of the partitioned GMM approach for complex, longitudinal datasets.

12.5 Joint Modeling for Correlated Binary Responses

Models such as those based on GMM are effective for addressing time-dependent covariates in longitudinal data. However, fitting separate models to multiple outcomes fails to account for the correlation between the outcome variables, leading to inefficient parameter estimates and potentially misleading conclusions, Berridge and Crouchley (2011) and Fitzmaurice (2004). While the multiple-equation GMM model, Hayashi (2000) allows for the simultaneous fit of multiple outcomes, it does not explicitly account for time-dependent covariates, limiting its applicability in longitudinal settings.

To address these shortcomings, we extend the partitioned GMM model (Irimata et al. 2019) to jointly model correlated binary outcomes, incorporating both time-dependent covariates and the dependencies between outcomes. This approach leverages simultaneous model fitting through an expanded objective function, improving the efficiency of regression parameter estimation and providing a more comprehensive understanding of the relationships in the data.

Consider a regression model with time-dependent covariates, and the method of Lalonde et al. (2014) to identify the set of valid moment conditions, though other methods for valid moment identification can be used. Let M denote the number of response variables of interest, where $\mathbf{Y}_{(m)i} = \left(y_{(m)i1}, \ldots, y_{(m)iT}\right)'$ is the $T \times 1$ vector of outcomes associated with the ith subject $i = 1, 2, \ldots, \mathrm{N}$; for the mth outcome variable and let $\boldsymbol{\mu}_{(m)i} = \left(\mu_{(m)i1}, \ldots, \mu_{(m)iT}\right)'$ denote the corresponding mean vector. Let $\mathbf{X}_{(m)i} = \begin{bmatrix} 1 & X_{(m)i11} & \cdots & X_{(m)iJ1} \\ \vdots & \vdots & \ddots & \vdots \\ \mathbf{1} & X_{(m)i1T} & \cdots & X_{(m)iJT} \end{bmatrix}$ denote the matrix of J covariates for model m, where at time t, the row vector is given by $\mathbf{X}_{(m)i.t} = \left(\mathrm{X}_{(m)i1t}, \ldots, X_{(m)iJt}\right)$ and the column vector for the jth covariate is $\mathbf{X}_{(m)ij.} = \left(\mathrm{X}_{(m)ij1}, \ldots, \mathrm{X}_{(m)ijT}\right)'$. To account for the time-dependent covariates and evaluate lagged effects, we utilize a lower triangular data matrix, similar to the method discussed by Irimata et al. (2019), such that the data matrix for the jth covariate in the mth model is

$$\mathbf{X}'_{(m)ij} = \begin{bmatrix} 1 & \mathrm{X}_{(\mathrm{m})\mathrm{ij}1} & 0 & \cdots & 0 \\ 1 & \mathrm{X}_{(m)ij2} & \mathrm{X}_{(m)ij1} & \cdots & 0 \\ \vdots & \vdots & \vdots & \cdots & \vdots \\ 1 & \mathrm{X}_{(\mathrm{m})\mathrm{ij}\mathrm{T}} & \mathrm{X}_{(m)ij(T-1)} & \cdots & \mathrm{X}_{(\mathrm{m})\mathrm{ij}1} \end{bmatrix}$$

$$= \begin{bmatrix} \mathbf{1} & \mathrm{X}^{[0]}_{(m)ij} & \mathrm{X}^{[1]}_{(m)ij} & \ldots & \mathrm{X}^{[T-1]}_{(m)ij} \end{bmatrix}$$

where the bracketed superscripts denote the differences in observed time periods between the mth outcome and the covariate. Thus, the model is

$$g\left(\mu_{(m)it}\right) = \beta_{(m)0} + \beta^{tt}_{(m)j}\mathrm{X}^{[0]}_{(m)ij} + \beta^{[1]}_{(m)j}\mathrm{X}^{[1]}_{(m)ij} + \beta^{[2]}_{(m)j}\mathrm{X}^{[2]}_{(m)ij} \cdots + \beta^{[T-1]}_{(m)\mathrm{j}}\mathrm{X}^{[T-1]}_{(m)ij}$$

for appropriate link function $g(\cdot)$, and $g\left(\boldsymbol{\mu}_{(m)i}\right) = \boldsymbol{X}_{(m)ij}\boldsymbol{\beta}_{(m)j}$ in matrix form. The interpretation of each of the regression coefficients is analogous to Model (12.2). Thus, $\beta^{tt}_{(m)j}$ denotes the cross-sectional effect of the jth covariate in the mth regression model, while $\beta^{[1]}_{(m)j}, \ldots, \beta^{[T-1]}_{(m)j}$ denote the lagged effects across a $1, \ldots, T-1$ time-period lag.

Let $M = 2$, with two simultaneous outcomes, which for subject i are the vectors $\mathbf{Y}_{(1)i}$ and $\mathbf{Y}_{(2)i}$, though the approach is analogous for larger M (Hayashi 2000). Let $\boldsymbol{\beta}_{(1)}$ and $\boldsymbol{\beta}_{(2)}$ be vectors of unknown regression parameters corresponding to the functions $\mathbf{f}\left(y_{(1)i},\boldsymbol{\beta}_{(1)}\right)=\begin{bmatrix} f_1\left(\mathrm{y}_{(1)i},\boldsymbol{\beta}_{(1)}\right) \\ \vdots \\ f_{G_1}\left(\mathrm{y}_{(1)i},\boldsymbol{\beta}_{(1)}\right) \end{bmatrix}$

and $\mathbf{h}\left(y_{(2)i},\boldsymbol{\beta}_{(2)}\right)=\begin{bmatrix} h_1\left(y_{(2)i},\boldsymbol{\beta}_{(2)}\right) \\ \vdots \\ h_{G_2}\left(y_{(2)i},\boldsymbol{\beta}_{(2)}\right) \end{bmatrix}$ for every subject i, where the number of moment conditions, G_1 and G_2 corresponding to each model may not be equal. Then, there are $G=\sum_{m=1}^{M} G_m$ moment conditions which take the form $\mathbf{f}\left(y_{(1)i},\boldsymbol{\beta}_{(1)}\right)$ and $\mathbf{h}\left(y_{(2)i},\boldsymbol{\beta}_{(2)}\right)$, such that

$$E\begin{bmatrix} \mathbf{f}\left(y_{(1)\mathrm{i}},\boldsymbol{\beta}_{(\mathbf{1})}\right) \\ \mathbf{h}\left(y_{(2)\mathrm{i}},\boldsymbol{\beta}_{(\mathbf{2})}\right) \end{bmatrix}=E\begin{bmatrix} f_1\left(y_{(1)\mathrm{i}},\boldsymbol{\beta}_{(\mathbf{1})}\right) \\ \vdots \\ f_{G_1}\left(y_{(1)\mathrm{i}},\boldsymbol{\beta}_{(\mathbf{1})}\right) \\ h_1\left(y_{(2)\mathrm{i}},\boldsymbol{\beta}_{(\mathbf{2})}\right) \\ \vdots \\ h_{G_2}\left(y_{(2)\mathrm{i}},\boldsymbol{\beta}_{(\mathbf{2})}\right) \end{bmatrix}=\begin{bmatrix} E\left(f_1\left(y_{(1)\mathrm{i}},\boldsymbol{\beta}_{(\mathbf{1})}\right)\right) \\ \vdots \\ E\left(f_{G_1}\left(y_{(1)\mathrm{i}},\boldsymbol{\beta}_{(\mathbf{1})}\right)\right) \\ E\left(h_1\left(y_{(2)\mathrm{i}},\boldsymbol{\beta}_{(\mathbf{2})}\right)\right) \\ \vdots \\ E\left(h_{G_2}\left(y_{(2)\mathrm{i}},\boldsymbol{\beta}_{(\mathbf{2})}\right)\right) \end{bmatrix}=\begin{pmatrix} 0 \\ \vdots \\ 0 \\ 0 \\ \vdots \\ 0 \end{pmatrix}$$

for all i. We define the sample analog of this moment condition across all subjects as

$$\begin{matrix} \mathbf{F}_{\mathbf{N}}\left(y_{(1)},\boldsymbol{\beta}_{(1)}\right) \\ \mathbf{H}_{\mathbf{N}}\left(y_{(2)},\boldsymbol{\beta}_{(2)}\right) \end{matrix}=\begin{matrix} \dfrac{1}{N}\displaystyle\sum_{i=1}^{N}\begin{bmatrix} f_1\left(y_{(1)i},\boldsymbol{\beta}_{(1)}\right) \\ \vdots \\ f_{G_1}\left(y_{(1)i},\boldsymbol{\beta}_{(1)}\right) \end{bmatrix} \\ \dfrac{1}{N}\displaystyle\sum_{i=1}^{N}\begin{bmatrix} h_1\left(y_{(2)i},\boldsymbol{\beta}_{(2)}\right) \\ \vdots \\ h_{G_2}\left(y_{(2)i},\boldsymbol{\beta}_{(2)}\right) \end{bmatrix} \end{matrix}=\frac{1}{N}\begin{bmatrix} \Sigma f_1\left(y_{(1)i},\boldsymbol{\beta}_{(1)}\right) \\ \vdots \\ \Sigma f_{G_1}\left(y_{(1)i},\boldsymbol{\beta}_{(1)}\right) \\ \Sigma h_1\left(y_{(2)i},\boldsymbol{\beta}_{(2)}\right) \\ \vdots \\ \Sigma h_{G_2}\left(y_{(2)i},\boldsymbol{\beta}_{(2)}\right) \end{bmatrix}$$

Define the $G \times G$ positive definite weight matrix as $W_N = \begin{pmatrix} W_{11} & W_{12} \\ W_{12} & W_{22} \end{pmatrix}$, which is computed using the cross product of the moment conditions for each subject. This matrix accounts for the covariance that exists between the two models through the inclusion of the elements in W_{12}. Thus,

$$W_N = \frac{1}{N}\left[\sum_{i=1}^{N}\begin{pmatrix} \mathbf{f}\left(\mathrm{y}_{(1)i}, \boldsymbol{\beta}_{(1)}\right) \\ \mathbf{h}\left(\mathrm{y}_{(2)i}, \boldsymbol{\beta}_{(2)}\right) \end{pmatrix}\begin{pmatrix} \mathbf{f}\left(\mathrm{y}_{(1)i}, \boldsymbol{\beta}_{(1)}\right) \\ \mathbf{h}\left(\mathrm{y}_{(2)i}, \boldsymbol{\beta}_{(2)}\right) \end{pmatrix}^T\right]$$

Define the objective function such that

$$Q_N(\mathbf{b}) = \begin{pmatrix} \mathbf{F}_N\left(y_{(1)}, \boldsymbol{\beta}_{(1)}\right) \\ \mathbf{H}_N\left(y_{(2)}, \boldsymbol{\beta}_{(2)}\right) \end{pmatrix}^T W_N^{-1} \begin{pmatrix} \mathbf{F}_N\left(y_{(1)}, \boldsymbol{\beta}_{(1)}\right) \\ \mathbf{H}_N\left(y_{(2)}, \boldsymbol{\beta}_{(2)}\right) \end{pmatrix}$$

The GMM estimator of $\boldsymbol{\beta} = \begin{pmatrix} \boldsymbol{\beta}_{(1)} \\ \boldsymbol{\beta}_{(2)} \end{pmatrix}$ is the vector of regression parameters that minimizes $Q_N(\boldsymbol{\beta})$, thus

$$\hat{\boldsymbol{\beta}}_{\text{GMM}} = \begin{pmatrix} \hat{\beta}_{\text{GMM}(1)} \\ \hat{\beta}_{\text{GMM}(2)} \end{pmatrix} = \operatorname{argmin}_{\beta \epsilon B} Q_N(\boldsymbol{\beta})$$

This minimum is obtained using nonlinear optimization, such as the Newton–Raphson or the conjugate gradient method. The estimates are obtained using a continuously updating procedure in which each successive estimate of $\boldsymbol{\beta}$ is obtained with a weight matrix calculated using the estimate for $\boldsymbol{\beta}$ from the previous iteration.

Let

$$\hat{\mathrm{A}} = \frac{1}{N}\sum_{i=1}^{N}\begin{pmatrix} \dfrac{\partial \mathrm{f}\left(y_{(1)i}, \hat{\beta}_{\text{GMM}(1)}\right)}{\partial \boldsymbol{\beta}_{(1)}} \\ \dfrac{\partial \mathrm{h}\left(y_{(2)i}, \hat{\beta}_{\text{GMM}(2)}\right)}{\partial \boldsymbol{\beta}_{(2)}} \end{pmatrix}$$

denote the vector of partial derivatives evaluated at $\hat{\boldsymbol{\beta}}_{\text{GMM}}$. The asymptotic variance of $\hat{\boldsymbol{\beta}}_{\text{GMM}}$ is $\text{Var}\left(\hat{\beta}_{GMM}\right)=\left(\hat{\text{A}}\text{W}_{\text{N}}^{-1}\hat{\text{A}}\right)^{-1}$, with W_{N}^{-1} evaluated at $\hat{\boldsymbol{\beta}}_{\text{GMM}}$.

By extension, for the general case $m = M$,

$$W_N = \frac{1}{N}\left[\sum_{i=1}^{N}\begin{pmatrix}\mathbf{f}\left(y_{(1)i},\boldsymbol{\beta}_{(1)}\right)\\ \mathbf{h}\left(y_{(2)i},\boldsymbol{\beta}_{(2)}\right)\\ \vdots \\ \mathbf{k}\left(y_{(m)i},\boldsymbol{\beta}_{(\mathbf{M})}\right)\end{pmatrix}\begin{pmatrix}\mathbf{f}\left(y_{(1)i},\boldsymbol{\beta}_{(1)}\right)\\ \mathbf{h}\left(y_{(2)i},\boldsymbol{\beta}_{(2)}\right)\\ \vdots \\ \mathbf{k}\left(y_{(m)i},\boldsymbol{\beta}_{(\mathbf{M})}\right)\end{pmatrix}^T\right], \text{ such that}$$

$$\text{Q}_{\text{N}}\left(\boldsymbol{\beta}\right)=\begin{pmatrix}\mathbf{F}_{\mathbf{N}}\left(y_{(1)},\boldsymbol{\beta}_{(1)}\right)\\ \mathbf{H}_{\mathbf{N}}\left(y_{(2)},\boldsymbol{\beta}_{(2)}\right)\\ \vdots \\ \mathbf{K}_{\mathbf{N}}\left(y_{(m)},\boldsymbol{\beta}_{(\mathbf{M})}\right)\end{pmatrix}^T \text{W}_{\text{N}}^{-1}\begin{pmatrix}\mathbf{F}_{\mathbf{N}}\left(y_{(1)},\boldsymbol{\beta}_{(1)}\right)\\ \mathbf{H}_{\mathbf{N}}\left(y_{(2)},\boldsymbol{\beta}_{(2)}\right)\\ \vdots \\ \mathbf{K}_{\mathbf{N}}\left(y_{(m)},\boldsymbol{\beta}_{(\mathbf{M})}\right)\end{pmatrix}$$

$$\text{and thus } \hat{\boldsymbol{\beta}}_{\mathbf{GMM}}=\begin{pmatrix}\hat{\boldsymbol{\beta}}_{\mathbf{GMM(1)}}\\ \hat{\boldsymbol{\beta}}_{\mathbf{GMM(2)}}\\ \vdots \\ \hat{\boldsymbol{\beta}}_{\mathbf{GMM}(\boldsymbol{M})}\end{pmatrix}=\text{argmin}_{\beta\epsilon B}\text{Q}_{\text{N}}\left(\boldsymbol{\beta}\right).$$

For logistic regression models on the mth outcome variable, the mean response is

$$\mu_{(m)it}\left(\boldsymbol{\beta}_{(\boldsymbol{m})}\right)=\frac{\exp\left(\mathbf{x}_{(m)it.}\boldsymbol{\beta}_{(\boldsymbol{m})}\right)}{1+\exp\left(\mathbf{x}_{(m)it.}\boldsymbol{\beta}_{(\boldsymbol{m})}\right)}$$

and thus, each element in $\mathbf{f}\left(y_{(1)i},\boldsymbol{\beta}_{(1)}\right)$ or $\mathbf{h}\left(y_{(2)i},\boldsymbol{\beta}_{(2)}\right)$ is

$$\frac{\partial \mu_{(m)is}\left(\boldsymbol{\beta}_{(\mathrm{m})}\right)}{\partial \boldsymbol{\beta}_{(m)j}^{[\mathrm{k}]}}\left[\mathrm{y}_{(\mathrm{m})\mathrm{it}}-\mu_{(\mathrm{m})\mathrm{it}}\left(\boldsymbol{\beta}_{(\mathrm{m})}\right)\right]=x_{(m)isj}\mu_{(m)is}\left(\boldsymbol{\beta}_{(m)}\right)\left[1-\mu_{(m)is}\left(\boldsymbol{\beta}_{(\mathrm{m})}\right)\right]\left[y_{(m)it}-\mu_{(m)it}\left(\boldsymbol{\beta}_{(m)}\right)\right]$$

Therefore, each row of $\begin{matrix}\dfrac{\partial f\left(y_{(1)\mathrm{i}},\hat{\boldsymbol{\beta}}_{\mathbf{GMM(1)}}\right)}{\partial \boldsymbol{\beta}_{(1)}}\\ \dfrac{\partial h\left(y_{(2)i},\hat{\boldsymbol{\beta}}_{\mathbf{GMM(2)}}\right)}{\partial \boldsymbol{\beta}_{(2)}}\end{matrix}$ in the asymptotic variance of a logistic regression model is computed using

$$\frac{\partial\left\{\left[\left[\dfrac{\partial \mu_{(m)is}\left(\boldsymbol{\beta}_{(m)}\right)}{\partial \beta_{(m)j}^{[k]}}\right]\left[y_{(m)it}-\mu_{(m)it}\left(\boldsymbol{\beta}_{(m)}\right)\right]\right]\right\}}{\partial \beta_{(m)l}^{[k']}}=$$

$$x_{(m)isj}\mu_{(m)is}\left(\boldsymbol{\beta}_{(m)}\right)\left[1-\mu_{(m)is}\left(\boldsymbol{\beta}_{(m)}\right)\right]$$
$$\left\{x_{(m)isl}\left[1-2\mu_{(\mathrm{m})\mathrm{is}}\left(\boldsymbol{\beta}_{(m)}\right)\right]\left[y_{(m)it}-\mu_{(m)it}\left(\boldsymbol{\beta}_{(m)}\right)\right]-x_{(m)itj}\mu_{(m)it}\left(\boldsymbol{\beta}_{(m)}\right)\left[1-\mu_{(m)it}\left(\boldsymbol{\beta}_{(m)}\right)\right]\right\}$$

where $j=1,\ldots,J$, $l=1,\ldots J$, $k=1,\ldots,T-1$ and $k'=1,\ldots,T-1$.

Analogously, for normal error models on the mth outcome variable, the elements in $\mathbf{f}\left(y_{(1)i},\boldsymbol{\beta}_{(1)}\right)$ or $\mathbf{h}\left(y_{(2)i},\boldsymbol{\beta}_{(2)}\right)$ are

$$\frac{\partial \mu_{(m)is}\left(\boldsymbol{\beta}_{(\mathrm{m})}\right)}{\partial \beta_{(m)\mathrm{j}}^{[\mathrm{k}]}}\left[y_{(m)it}-\mu_{(m)it}\left(\boldsymbol{\beta}_{(m)}\right)\right]=x_{(m)isj}\left[y_{(m)it}-\mu_{(m)it}\left(\boldsymbol{\beta}_{(m)}\right)\right]$$

and thus, in calculating the asymptotic variance of $\hat{\boldsymbol{\beta}}_{\mathrm{GMM}}$, each row of $\begin{pmatrix}\dfrac{\partial f\left(y_{(1)i},\hat{\boldsymbol{\beta}}_{GMM(1)}\right)}{\partial \boldsymbol{\beta}_{(1)}}\\ \dfrac{\partial h\left(y_{(2)i},\hat{\boldsymbol{\beta}}_{GMM(2)}\right)}{\partial \boldsymbol{\beta}_{(2)}}\end{pmatrix}$ is computed as

$$\frac{\partial\left\{\left[\frac{\partial\mu_{(m)is}\left(\boldsymbol{\beta}_{(m)}\right)}{\partial\beta_{(m)j}^{[k]}}\right]\left[y_{(m)it}-\mu_{(m)it}\left(\boldsymbol{\beta}_{(m)}\right)\right]\right\}}{\partial\beta_{(m)l}^{[k']}}=-x_{(m)isj}x_{(m)isl}$$

for $j=1,\ldots,J$, $l=1,\ldots J$, $k=1,\ldots,T-1$ and $k'=1,\ldots,T-1$. We provide a SAS macro to fit this model at https://github.com/kirimata/Simultaneous-GMM.

12.6 GMM Estimation for Time-Dependent Covariate Models

Consider a regression model with time-dependent covariates. To identify valid moment conditions, we employ the method proposed by Lalonde et al. (2014), though other methods can also be applied. Let M denote the number of response variables of interest, where $\mathbf{Y}_{(\mathrm{m})\mathrm{i}}=(y_{(m)i1},\ldots,y_{(m)iT})'$ represents the $T\times 1$ vector of outcomes for subject $i=1,2,\ldots,N$ for the mth outcome variable. Let $\boldsymbol{\mu}_{(m)i}=\left(\mu_{(m)i1},\ldots,\mu_{(m)iT}\right)'$ denote the corresponding mean vector.

Define Let $\mathbf{X}_{(m)i}=\begin{bmatrix}1 & X_{(m)i11} & \cdots & X_{(m)iJ1}\\ \vdots & \vdots & \ddots & \vdots\\ 1 & X_{(m)i1T} & \cdots & X_{(m)iJT}\end{bmatrix}$.

At time t, the row vector is $\mathbf{X}_{(m)i.t}=\left(\mathrm{X}_{(m)i1t},\ldots,X_{(m)iJt}\right)$, and the column vector for the jth covariate is $\mathbf{X}_{(\boldsymbol{m})\boldsymbol{ij}}=(X_{(m)ij1},\ldots,X_{(m)ijT})'$. To model time-dependent covariates and assess lagged effects, we utilize a lower triangular structure for the data matrix, similar to the approach discussed by Irimata et al. (2018):

$$\mathbf{X}'_{(m)ij}=\begin{bmatrix}1 & \mathrm{X}_{(\mathrm{m})\mathrm{ij}1} & 0 & \cdots & 0\\ 1 & X_{(m)ij2} & X_{(m)ij1} & \cdots & 0\\ \vdots & \vdots & \vdots & \cdots & \vdots\\ 1 & \mathrm{X}_{(\mathrm{m})\mathrm{ij}\mathrm{T}} & X_{(m)ij(T-1)} & \cdots & \mathrm{X}_{(\mathrm{m})\mathrm{ij}1}\end{bmatrix}=\begin{bmatrix}\mathbf{1} & X_{(m)ij}^{[0]} & X_{(m)ij}^{[1]} & \cdots & X_{(m)ij}^{[T-1]}\end{bmatrix}$$

where the superscripts in brackets denote the time lags. The regression model is then expressed as

$$g\left(\mu_{(m)it}\right)=\beta_{(m)0}+\beta^{tt}_{(m)j}X^{[0]}_{(m)ij}+\beta^{[1]}_{(m)j}X^{[1]}_{(m)ij}+\beta^{[2]}_{(m)j}X^{[2]}_{(m)ij}\cdots+\beta^{[T-1]}_{(\mathrm{m})\mathrm{j}}X^{[T-1]}_{(m)ij}$$

or in matrix form:

$$\mathrm{g}\left(\boldsymbol{\mu}_{(\mathbf{m})\mathbf{i}}\right)=\mathbf{X}_{(\mathbf{m})\mathbf{ij}}\boldsymbol{\beta}_{(\mathbf{m})\mathbf{j}}$$

where $g(\cdot)$ is an appropriate link function. The coefficients $\beta^{[1]}_{(m)j},\ldots,\beta^{[T-1]}_{(m)j}$ represent the cross-sectional and lagged effects, respectively.

12.6.1 GEEs for Multiple Outcomes

Now let $M = 2$, where $\mathbf{Y}_{(1)i}$ and $\mathbf{Y}_{(2)i}$ represent the simultaneous outcomes for subject iii. Define the regression parameter vectors $\boldsymbol{\beta}_{(1)}$ and $\boldsymbol{\beta}_{(2)}$ for the functions:

$$\mathbf{f}\left(y_{(1)i},\boldsymbol{\beta}_{(1)}\right)=\begin{bmatrix} f_1\left(y_{(1)i},\boldsymbol{\beta}_{(1)}\right)\\ \vdots \\ f_{G_1}\left(y_{(1)i},\boldsymbol{\beta}_{(1)}\right)\end{bmatrix} \text{ and } \mathbf{h}\left(y_{(2)I},\boldsymbol{\beta}_{(2)}\right)=\begin{bmatrix} h_1\left(y_{(2)I},\boldsymbol{\beta}_{(2)}\right)\\ \vdots \\ h_{G_2}\left(y_{(2)I},\boldsymbol{\beta}_{(2)}\right)\end{bmatrix}$$

The total number of moment conditions is $G = G_1 + G_2$. The expectations of the moment conditions are

$$E\begin{bmatrix}\mathbf{f}\left(y_{(1)i},\boldsymbol{\beta}_{(1)}\right)\\ \mathbf{h}\left(y_{(2)i},\boldsymbol{\beta}_{(2)}\right)\end{bmatrix}=E\begin{bmatrix} f_1\left(y_{(1)i},\boldsymbol{\beta}_{(1)}\right)\\ \vdots \\ f_{G_1}\left(y_{(1)i},\boldsymbol{\beta}_{(1)}\right)\\ h_1\left(y_{(2)i},\boldsymbol{\beta}_{(2)}\right)\\ \vdots \\ h_{G_2}\left(y_{(2)i},\boldsymbol{\beta}_{(2)}\right)\end{bmatrix}=\begin{bmatrix} E\left(f_1\left(y_{(1)i},\boldsymbol{\beta}_{(1)}\right)\right)\\ \vdots \\ E\left(f_{G_1}\left(y_{(1)i},\boldsymbol{\beta}_{(1)}\right)\right)\\ E\left(h_1\left(y_{(2)i},\boldsymbol{\beta}_{(2)}\right)\right)\\ \vdots \\ E\left(h_{G_2}\left(y_{(2)i},\boldsymbol{\beta}_{(2)}\right)\right)\end{bmatrix}=\begin{pmatrix}0\\ \vdots \\ 0\\ 0\\ \vdots \\ 0\end{pmatrix}$$

The sample analog of this condition is

$$\begin{matrix} \mathbf{F}_N\left(y_{(1)}, \boldsymbol{\beta}_{(1)}\right) \\ \mathbf{H}_N\left(y_{(2)}, \boldsymbol{\beta}_{(2)}\right) \end{matrix} = \begin{matrix} \frac{1}{N}\sum_{i=1}^{N}\begin{bmatrix} f_1\left(y_{(1)i}, \boldsymbol{\beta}_{(1)}\right) \\ \vdots \\ f_{G_1}\left(y_{(1)i}, \boldsymbol{\beta}_{(1)}\right) \end{bmatrix} \\ \frac{1}{N}\sum_{i=1}^{N}\begin{bmatrix} h_1\left(y_{(2)i}, \boldsymbol{\beta}_{(2)}\right) \\ \vdots \\ h_{G_2}\left(y_{(2)i}, \boldsymbol{\beta}_{(2)}\right) \end{bmatrix} \end{matrix} = \frac{1}{N}\begin{bmatrix} \Sigma f_1\left(y_{(1)i}, \boldsymbol{\beta}_{(1)}\right) \\ \vdots \\ \Sigma f_{G_1}\left(y_{(1)i}, \boldsymbol{\beta}_{(1)}\right) \\ \Sigma h_1\left(y_{(2)i}, \boldsymbol{\beta}_{(2)}\right) \\ \vdots \\ \Sigma h_{G_2}\left(y_{(2)i}, \boldsymbol{\beta}_{(2)}\right) \end{bmatrix}$$

Define the weight matrix W_N as $W_N = \begin{pmatrix} W_{11} & W_{12} \\ W_{12} & W_{22} \end{pmatrix}$, The GMM estimator minimizes the objective function:

$$Q_N(\boldsymbol{\beta}) = \begin{pmatrix} \mathbf{F}_N\left(y_{(1)}, \boldsymbol{\beta}_{(1)}\right) \\ \mathbf{H}_N\left(y_{(2)}, \boldsymbol{\beta}_{(2)}\right) \end{pmatrix}^T \mathrm{W}_N^{-1} \begin{pmatrix} \mathbf{F}_N\left(y_{(1)}, \boldsymbol{\beta}_{(1)}\right) \\ \mathbf{H}_N\left(y_{(2)}, \boldsymbol{\beta}_{(2)}\right) \end{pmatrix}$$

The estimate is

$$\boldsymbol{\beta}_{\mathrm{GMM}} = \begin{pmatrix} \hat{\beta}_{\mathbf{GMM(1)}} \\ \hat{\beta}_{\mathbf{GMM(2)}} \end{pmatrix} = \operatorname{argmin}_{\beta \epsilon B} Q_N(\boldsymbol{\beta})$$

12.6.2 Asymptotic Variance

The asymptotic variance is computed as

$$\mathrm{Var}\left(\hat{\boldsymbol{\beta}}_{\mathrm{GMM}}\right) = \left(\hat{A}\mathrm{W}_N^{-1}\hat{A}\right)^{-1},$$

where

$$\hat{A} = \frac{1}{N}\sum_{i=1}^{N}\begin{pmatrix} \dfrac{\partial \mathrm{f}\left(y_{(1)i}, \hat{\boldsymbol{\beta}}_{\mathbf{GMM(1)}}\right)}{\partial \boldsymbol{\beta}_{(1)}} \\ \dfrac{\partial h\left(y_{(2)i}, \hat{\boldsymbol{\beta}}_{\mathbf{GMM(2)}}\right)}{\partial \boldsymbol{\beta}_{(2)}} \end{pmatrix}$$

The same approach can be extended to M-dimensional outcomes. For additional details and implementation, a SAS macro is available at this GitHub repository.

12.7 Simulation Study

We implemented a simulation study in order to investigate the performance of the Simultaneous GMM, as compared to the use of separate models. This simulation expanded on the approach discussed by Irimata et al. (2019). Data were simulated for $T=3$ time periods, with $N=250$, and 500 subjects with two response variables. Responses were simulated from a Bernoulli distribution, with the marginal mean depending on a single continuous Type II time-dependent covariate. This predictor had a randomly generated time-dependent covariate weight between 2 and 4 across a one time-period lag, and a slightly weaker randomly generated weight between 1 and 2 across a two time-period lag. The clustering effect representing a longitudinal survey was incorporated using a random effect following a normal distribution with a mean of 0 and variance of 1. Correlation between the two responses was incorporated using a second random effect following a normal distribution with mean 0 and variance between 0 and 4. Thus, we investigated the effect of varying amounts of correlation between the response variables. The lagged regression parameters for the first model were assigned to be $\beta_{(1)0}=0$, $\beta^{tt}_{(1)}=0.3$, $\beta^{[1]}_{(1)}=0.7$ and $\beta^{[2]}_{(1)}=0.5$, and the regression parameters for the second model were set to $\beta_{(2)0}=0$, $\beta^{tt}_{(2)}=0.3$, $\beta^{[1]}_{(2)}=0.3$, and $\beta^{[2]}_{(2)}=0.7$.

Data were simulated for $T = 3$ time periods, with sample sizes of $N = 250$ and $N = 500$, and included two binary response variables. The responses were generated from a Bernoulli distribution, with marginal means dependent on a single continuous Type II time-dependent covariate. This predictor featured a randomly generated time-dependent covariate weight between 2 and 4 across a one-time-period lag and a slightly weaker weight between 1 and 2 across a two-time-period lag.

To incorporate a longitudinal survey structure, we included a random effect to account for clustering within subjects. This random effect followed a normal distribution with a mean of 0 and variance of 1. Additionally, correlation between the two response variables was introduced through a second random effect, also following a normal distribution with a mean of 0 but with variance varying between 0 and 4 to represent different levels of correlation between the responses. This allowed us to investigate how varying levels of correlation impacted the models' performance.

The regression parameters for the first model were set as follows: $\beta_{(1)0} = 0$, $\beta^{tt}_{(1)} = 0.3$, $\beta^{[1]}_{(1)} = 0.7$, and $\beta^{[2]}_{(1)} = 0.5$, and the regression parameters for the second model were set to $\beta_{(2)0} = 0$, $\beta^{tt}_{(2)} = 0.3$, $\beta^{[1]}_{(2)} = 0.3$, and $\beta^{[2]}_{(2)} = 0.7$.

Through this simulation, we aimed to assess the ability of the simultaneous GMM approach to effectively estimate model parameters and handle correlations between response variables, especially in the presence of lagged covariate effects and varying levels of interresponse correlation.

12.8 Numerical Example

Alcohol use and smoking have long been identified as strongly related Guydish et al. (2011) and Kozlowski et al. (1986). Recent studies, such as the National Comorbidity Study, have reported smoking prevalence of 56.1% for Americans with alcohol disorders (Lasser et al. 2000). We analyzed data from the National Longitudinal Study of Adolescent Health (Add Health) (Harris and Udry 2022) which is a longitudinal study of health-related behaviors in adolescents. The measurements were collected from 80 different high schools and 52 middle schools in the United States, with information on 2,712 different students gathered across four waves starting in 1994 and ending in 2008. We investigated two binary outcomes representing smoking status and social alcohol drinking at each wave. The data included time-dependent covariates for physical activity level, depression level, and self-reported health status. We also included a time-independent covariate, representing race as white or nonwhite. For further comparison, we considered a third binary outcome representing obesity status, as calculated from each child's BMI.

12.8.1 Simultaneous Modeling of Smoking and Social Alcohol Use

We considered two outcome variables for smoking status and social alcohol drinking, thus for this analysis $M = 2$. There was substantial association between these two outcome variables, with correlation $\hat{\phi} = 0.40$, and with $V^2 = \chi^2/N = 0.161$. Thus, since V^2 exceeded the 0.15 threshold discussed by Irimata and Wilson (2018), there was significant correlation between these outcomes, warranting consideration of this association in the model. We fitted a simultaneous GMM model to estimate the regression for the two outcome variables simultaneously. In addition, we considered two partitioned GMM models separately, using the Lalonde et al. (2014) method to identify valid moment conditions. There were differences in the significance of the covariates between these two models. In particular, for the social alcohol usage model, we saw that depression had a significant one-period lagged coefficient when two separate models were fit. However, this relationship was not significant when the correlation between smoking and social alcohol usage was accounted for through the simultaneous GMM. We found that depression level was significant at a two time-period lag under the simultaneous GMM model, though the two separate models did not identify this relationship as significant. Thus, strong correlation between smoking status and social alcohol drinking affects the conclusions from a binary logistic regression model. In particular, the fit of two separate models is not the best choice in this scenario as it neglects the association between the outcomes, consistent with findings based on the correlation of the two outcomes (Irimata et al. 2019). The parameter estimates and standard errors from the simultaneous GMM model, as well as the two-separate partitioned GMM models are included in Table 12.1. Significant parameter estimates are denoted by bolded entries.

The results highlighted important differences in the significance of covariates between the two modeling approaches:

- **Depression:** In the separate models, depression had a significant one-period lagged effect in the social alcohol use model. However, when the correlation between smoking and social alcohol use was accounted for in the simultaneous GMM, this relationship was no longer significant.
- **Lagged Effects:** In contrast, the simultaneous GMM revealed a significant two-period lagged effect of depression in the social alcohol use model that was not detected in the separate models.

These findings emphasize that the correlation between smoking and social alcohol drinking can influence conclusions drawn from binary logistic regression models. The simultaneous GMM approach accounts for this correlation, providing a more robust framework for analysis. Parameter estimates and standard errors from both the simultaneous and separate models are presented in Table 12.1.

12.8.2 Simultaneous Modeling of Smoking, Social Alcohol Use, and Obesity

To extend the analysis, we included a third binary outcome, **obesity status**, alongside smoking status and social alcohol use. Partial correlations were computed among the three outcomes:

- Smoking and social alcohol use remained strongly correlated ($\hat{\phi} = 0.40$, $V^2 = 0.161$).
- Neither smoking nor social alcohol use had strong correlations with obesity ($V^2 = 0.0041$, respectively).

TABLE 12.1

Parameter Estimates and Standard Errors (SE) for the Smoking and Social Alcohol Models in Add Health Data (Bolded = statistically significant with p-value < 0.05)

			Simultaneous GMM		Separate Models	
Outcome	**Time period**	**Parameter**	**Estimate**	**SE**	**Estimate**	**SE**
Smoking status	Cross-sectional	**Intercept**	**0.849**	**0.152**	**0.824**	**0.155**
		Race	**−0.757**	**0.068**	**−0.716**	**0.068**
		Activity	−0.023	0.022	−0.019	0.022
		Depression	**0.806**	**0.086**	**0.811**	**0.088**
		Health	**−0.259**	**0.030**	**−0.260**	**0.030**
	Lagged one period	**Activity**	−0.002	0.022	0.007	0.022
		Depression	0.002	0.089	0.018	0.091
		Health	**−0.089**	**0.023**	**−0.101**	**0.024**
	Lagged two periods	**Activity**	**0.063**	**0.023**	**0.063**	**0.024**
		Depression	**1.154**	**0.101**	**1.171**	**0.101**
	Lagged three periods	**Activity**	−0.002	0.028	0.004	0.028
		Depression	**−0.587**	**0.094**	**−0.606**	**0.094**
Social alcohol use	Cross-sectional	**Intercept**	0.208	0.156	0.207	0.159
		Race	**−0.562**	**0.065**	**−0.555**	**0.066**
		Activity	**−0.067**	**0.023**	**−0.071**	**0.023**
		Depression	**0.864**	**0.101**	**0.877**	**0.103**
		Health	**−0.110**	**0.029**	**−0.110**	**0.030**
	Lagged one period	**Activity**	0.009	0.023	−0.002	0.023
		Depression	−0.207	0.114	**−0.257**	**0.113**
		Health	0.004	0.024	0.016	0.024
	Lagged two periods	**Activity**	0.039	0.032	0.033	0.033
		Depression	**0.384**	**0.167**	0.253	0.155
		Health	**0.199**	**0.039**	**0.228**	**0.039**
	Lagged three periods	**Depression**	−0.204	0.186	−0.313	0.177
		Health	**0.134**	**0.040**	**0.164**	**0.041**

TABLE 12.2

Partial Correlations/ V^2 between Smoking, Social Alcohol Usage, and Obesity in Add Health Data

	Smoking	**Alcohol**	**Obesity**
Smoking	1	0.399/0.161	0.040/0.0041
Alcohol	0.399/0.161	1	0.046/0.0046
Obesity	0.040/0.0041	0.046/0.0046	1

These results suggest that including obesity in the simultaneous model would have little impact on the relationship between smoking and social alcohol use, given the weak correlation between obesity and the other outcomes. Partial correlations and V^2 values are shown in Table 12.2.

Using the three-outcome simultaneous GMM, we compared the results to those obtained from three separate Partitioned GMM models:

- **Social Alcohol Use:** Similar to the two-outcome model, depression was significant at a one-period lag in the separate models but not in the simultaneous model. Conversely, the simultaneous GMM identified depression as significant at a two-period lag, which was not detected in the separate models.

TABLE 12.3

Parameter Estimates and Standard Errors (SE) for the Smoking, Social Alcohol, and Obesity Models in Add Health Data (Bolded= statistically significant at p-value <0.05)

			Simultaneous GMM		Separate Models	
Outcome	**Time Period**	**Parameter**	**Estimate**	**SE**	**Estimate**	**SE**
Smoking status	Cross-sectional	**Intercept**	**0.823**	**0.148**	**0.824**	**0.155**
		Race	**−0.791**	**0.067**	**−0.716**	**0.068**
		Activity	−0.021	0.021	−0.019	0.022
		Depression	**0.687**	**0.082**	**0.811**	**0.088**
		Health	**−0.240**	**0.029**	**−0.260**	**0.030**
	Lagged one period	**Activity**	0.001	0.022	0.007	0.022
		Depression	−0.051	0.086	0.018	0.091
		Health	**−0.076**	**0.023**	**−0.101**	**0.024**
	Lagged two periods	**Activity**	**0.075**	**0.023**	**0.063**	**0.024**
		Depression	**1.181**	**0.098**	**1.171**	**0.101**
	Lagged three periods	**Activity**	0.032	0.028	0.004	0.028
		Depression	**−0.704**	**0.092**	**−0.606**	**0.094**
Social alcohol use	Cross-sectional	**Intercept**	0.271	0.154	0.207	0.159
		Race	**−0.586**	**0.065**	**−0.555**	**0.066**
		Activity	**−0.062**	**0.023**	**−0.071**	**0.023**
		Depression	**0.794**	**0.097**	**0.877**	**0.103**
		Health	**−0.120**	**0.029**	**−0.110**	**0.030**
	Lagged one period	**Activity**	0.035	0.023	−0.002	0.023
		Depression	−0.209	0.111	**−0.257**	**0.113**
		Health	−0.008	0.023	0.016	0.024
	Lagged two periods	**Activity**	0.032	0.032	0.033	0.033
		Depression	**0.331**	**0.156**	0.253	0.155
		Health	**0.217**	**0.038**	**0.228**	**0.039**
	Lagged three periods	**Depression**	−0.244	0.179	−0.313	0.177
		Health	**0.154**	**0.040**	**0.164**	**0.041**
Obesity	Cross-sectional	**Intercept**	−0.252	0.230	−0.104	0.228
		Race	**0.415**	**0.103**	**0.361**	**0.103**
		Activity	**−0.567**	**0.054**	**−0.566**	**0.054**
		Depression	**0.562**	**0.118**	**0.589**	**0.117**
		Health	**−0.526**	**0.047**	**−0.559**	**0.047**
	Lagged one period	**Depression**	**1.553**	**0.146**	**1.469**	**0.143**
		Health	**−0.201**	**0.042**	**−0.183**	**0.041**

- **Obesity:** The predictors for obesity were consistent across both the simultaneous and separate models, reflecting the weak correlation between obesity and the other outcomes.

These findings reinforce the need for simultaneous modeling when outcomes are correlated, as separate models may overlook important relationships or incorrectly identify others. Parameter estimates and standard errors for all three outcomes are provided in Table 12.3, with significant parameters bolded.

12.9 Discussions and Comments

This chapter demonstrates that simultaneous GMM modeling effectively accounts for correlations between outcomes, such as smoking and social alcohol use, providing more reliable parameter estimates than separate models. While the inclusion of obesity had minimal impact on the results due to its weak correlation with smoking and alcohol use, the observed discrepancies in the significance of depression effects highlight the value of simultaneous modeling for correlated outcomes. These findings are consistent with previous literature (Irimata et al. 2019) and underscore the importance of addressing outcome correlations in longitudinal health behavior studies.

In longitudinal data analysis, various types of correlation must be considered. These include correlation arising from repeated measures over time, as well as correlation between covariates and responses. When analyzing multiple outcome variables, it is crucial to account for the association between these outcomes as well. The simultaneous GMM model offers an effective alternative to traditional joint modeling of logistic regression with time-dependent covariates, as it avoids the need for additional distributional assumptions.

The simultaneous GMM model handles time-dependent covariates by employing partitioned coefficients, which capture the effect of each predictor at different time segments. Additionally, the model accounts for the correlation between multiple outcomes by simultaneously minimizing an extended weight matrix, which incorporates concatenated moment conditions. This approach provides a more robust understanding of the relationships between outcomes and covariates over time.

Looking forward, future research could explore extending this model to explicitly incorporate feedback loops, where the outcomes influence the covariates. This would offer a more comprehensive understanding of dynamic interactions in longitudinal data.

Advantages of the Joint Model:

1. **Improved Efficiency**: By simultaneously estimating parameters for all outcomes, the model leverages shared information across outcomes and time points, reducing variance and improving the precision of estimates.
2. **Accounting for Correlation**: The joint modeling approach directly incorporates the correlation between outcomes, addressing a key limitation of separate model fitting.
3. **Time-Dependent Covariates**: By building on the partitioned GMM framework, the model captures the dynamic relationships between covariates and outcomes across time, offering a more realistic representation of longitudinal data.

Applications and Extensions:
This joint modeling framework is particularly well-suited for studies involving multivariate binary outcomes observed over time, such as in clinical trials, public health studies, or educational research. Future extensions could explore nonlinear link functions, higher-order interactions, or alternative methods for specifying the correlation structure (e.g., random effects or hierarchical modeling approaches).

By combining the strengths of partitioned GMM and joint modeling techniques, this approach addresses the critical need for methods that accommodate both time-dependent covariates and correlated outcomes in longitudinal studies.

References

Bandyopadhyay, S., Ganguli, B., and Chatterjee, A. (2011). A review of multivariate longitudinal data analysis. *Statistical Methods in Medical Research, 20*(4), 299–330.

Berridge, D. M. and Crouchley, R. (2011). *Multivariate Generalized Linear Mixed Models Using R*. Boca Raton, FL: CRC Press.

Chen, I. C. and Westgate, P. M. (2017). Improved methods for the marginal analysis of longitudinal data in the presence of time-dependent covariates. *Statistics in Medicine, 36*(16), 2533–2546.

Fieuws, S. and Verbeke, G. (2004). Joint modelling of multivariate longitudinal profiles: pitfalls of the random-effects approach. *Statistics in Medicine, 23*(20), 3093–104.

Fieuws, S., Verbeke, G., and Molenberghs, G. (2007). Random-effects models for multivariate repeated measures. *Statistical Methods in Medical Research, 16*(5), 387–397.

Fitzmaurice, G. M., Laird, N. M., and Ware, J. H. (2004). *Applied Longitudinal Analysis*. Hoboken, NJ: Wiley.

Ghebremichael, M. (2015). Joint modeling of correlated binary outcomes: HIV-1 and HSV-2 co-infection. *Journal of Applied Statistics, 42*(10), 2180–2191.

Gueorguieva, R. (2001). A multivariate generalized linear mixed model for joint modelling of clustered outcomes in the exponential family. *Statistical Modelling, 1*(3), 177–193.

Guydish, J., Passalacqua, E., Tajima, B., Chan, M., Chun, J., and Bostrom, A. (2011). Smoking prevalence in addiction treatment: A review. *Nicotine & Tobacco Research, 13*(6), 401–411.

Hansen, L. P. (1982). Large sample properties of generalized method of moments estimators. *Econometrica, 50*(4), 1029–1054.

Harris, K. M. and Udry, J. R. (2022). National Longitudinal Study of Adolescent to Adult Health (Add Health), 1994-2018 [Public Use]. Carolina Population Center, University of North Carolina-Chapel Hill [distributor], Inter-university Consortium for Political and Social Research, https://doi.org/10.3886/ICPSR21600.v25

Hayashi, F. (2000). *Econometrics*. Princeton, NJ: Princeton University Press.

Hu, F. B., Goldberg, J., Hedeker, D., Flay, B. R., and Pentz, M. A. (1998). Comparison of population-averaged and subject-specific approaches for analyzing repeated binary outcomes. *American Journal of Epidemiology, 147*(7), 694–703.

Irimata, K. M., Broatch, J., and Wilson, J. R. (2019). Partitioned GMM logistic regression models for longitudinal data. *Statistics in Medicine, 38*(12), 2171–2183.

Irimata, K. M. and Wilson, J. R. (2018). Identifying intraclass correlation necessitating hierarchical modeling. *Journal of Applied Statistics, 45*(4), 626–641.

Kozlowski, L. T., Jelinek, L. C., and Pope, M. A. (1986). Cigarette smoking among alcohol abusers: A continuing and neglected problem. *Canadian Journal of Public Health, 77*(3), 205–207.

Lai, T. L. and Small, D. (2007). Marginal regression analysis of longitudinal data. *Journal of the Royal Statistical Society Series B (Statistical Methodology), 69*(1), 79–99.

Lalonde, T. L., Wilson, J. R., and Yin, J. (2014). GMM logistic regression models for longitudinal data with time-dependent covariates and extended classifications. *Statistics in Medicine, 33*(27), 4756–4769. doi:10.1002/sim.6273

Lasser, K., Boyd, J. W., Woolhandler, S., Himmelstein, D. U., McCormick, D., and Bor, D. H. (2000). Smoking and mental illness: A population-based prevalence study. *Journal of the American Medical Association, 284*(20), 2606–2610.

Liu, L., Strawderman, R. L., Cowen, M. E., and Shih, Y. T. (2010). A flexible two-part random effects model for correlated medical costs. *Journal of Health Economics, 29*(1), 110–123.

McCulloch, C. (2008). Joint modelling of mixed outcome types using latent variables. *Statistical Methods in Medical Research, 17*(1), 53–73.

Muller, H-G. and Stadtmuller, U. (2005). Generalized functional linear models. *Annals of Statistics, 33*(2), 774–805.

Pepe, M. S. and Anderson, G. L. (1994). A cautionary note on inference for marginal regression models with longitudinal data and general correlated response data. *Communications in Statistics – Simulation and Computation, 23*(4), 939–951.

Proust-Lima, C., Sene, M., Taylor, J. M., and Jacqmin-Gadda, H. (2014). Joint latent class models for longitudinal and time-to-event data: A review. *Statistical Methods in Medical Research, 23*(1), 74–90.

Qu, A., Lindsay, B. G., and Li, B. (2000). Improving generalized estimating equations using quadratic inference functions. *Biometrika, 87*(4), 823–836.

Song, X., Davidian, M., and Tsiatis, A. A. (2002). A semiparametric likelihood approach to joint modeling of longitudinal and time-to-event data. *Biometrics. 58*(4), 742–753.

Stoner, J. A., Leroux, B. G., and Puumala, M. (2010). Optimal combination of estimating equations in the analysis of multilevel nested correlated data. *Statistics in Medicine, 29*(4), 464–473.

Wu, L., Liu, W., Yi, G. Y., and Huang, Y. (2012). Analysis of longitudinal and survival data: Joint modeling, inference methods, and issues. *Journal of Probability and Statistics. 2012*, 1, 640153.

Zeger, S. L. and Liang, K. Y. (1986) Longitudinal data analysis for discrete and continuous outcomes. *Biometrics, 42*(1), 121–130.

Zhou, Y., Lefante, J., Rice, J., and Chen, S. (2014). Using modified approaches on marginal regression analysis of longitudinal data with time-dependent covariates. *Statistics in Medicine, 33*(19), 3354–3364.

13

Multiple Membership Mixed-Effects Models

13.1 Research Interest/Question

It is natural in a healthcare system to find that each patient receives their healthcare needs from more than one doctor. It is more often than not that each patients will get their healthcare from two or more doctors. Such relation leads to a hierarchical structure referred to as multiple memberships in multilevel statistical modeling. This relationship results in a hierarchical structure known as multiple membership in multilevel statistical modeling.

If a patient sees only one doctor, then this patient is completely nested within that doctor. Such case is referred to as single membership in multilevel statistical modeling as seen in Figure 13.1, which is the case discussed in Chapter 10 (Wilson et al. 2023). However, when a patient receives care from two or more doctors, which is more common in real-life healthcare systems. This patient is no longer completely nested within a doctor but nested within multiple doctors. Such a scenario is referred to as multiple membership as illustrated in Figure 13.1. As seen in Figure 13.1, the patient 1 received care from the doctor 1 in single membership as shown in the left panel, but received care from both doctors 1 and 2 as shown in the right panel, which leads to multiple membership.

In practice, multiple membership is more common and therefore requires careful attention in statistical modeling. This chapter is designed to extend the classical single membership multilevel modeling in Chapter 10 to illustrate how to analyze multiple membership multilevel structured data using SAS and R.

DOI: 10.1201/9781003596370-2

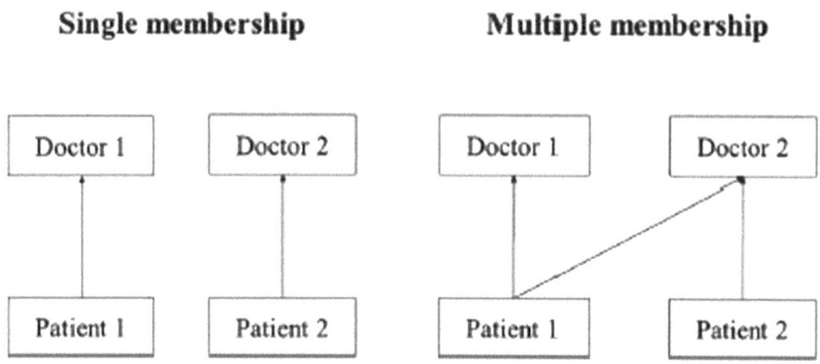

FIGURE 13.1
Single Membership Versus Multiple Membership Nesting Structure.

13.2 Hospital Doctor Patient Data with Multiple Membership

Recall the hospital doctor patient (HDP) data, as used in Chapters 4, 5, 6, and 10. The original HDP was designed for a three-level data structure where patients are nested within doctors, and doctors are nested within hospitals. The hierarchical structures at the three levels are single membership structured, as we assumed each patient is completely nested within doctor (visit for their health care needs), and each doctor is completely nested within hospital. The original hierarchical structure consists of 8,525 patients nested within 407 doctors, and these 407 doctors are nested within 35 hospitals. The nesting structure of these patients, doctors, and hospitals are illustrated in Table 10.2 and Table 10.3 in Wilson et al. (2023).

For this illustration of the multiple membership multilevel modeling, a set of multiple memberships based on the original HDP data was generated. To facilitate multiple membership, patients were randomly assigned to 2, 3, 4, 5, and 6 doctors with probabilities of 25%, 20%, 15%, 10%, and 5%, respectfully. Table 13.1 shows the multiple membership nesting structure for the first ten patients from the first hospital who are designed to be seen by the primary doctor (denoted by "cD1") along with other doctors in the hospital (denoted by "cD2" to "cD6" since the maximum number of doctors they can see is 6). As seen from Table 13.1 (left panel), the first patient (PID #1) visited five doctors as denoted by doctors #1 (in column "cD1"), doctor #82 (in column "cD2"), doctor #158 (in column "cD3"), doctor #229 (in column "cD4"), and doctor #298 (in column "cD5") for their health needs. Similarly, the third patient (PID #3) visited six doctors as denoted by doctors #1 (in column "cD1"), doctor #15 (in column "cD2"), doctor #136 (in column "cD3"), doctor #153 (in column "cD4"), doctor #174 (in column "cD5"), and doctor #294 (in

TABLE 13.1

A Subset of Data for Illustration

PID	cD1	cD2	cD3	cD4	cD5	cD6	cW1	cW2	cW3	cW4	cW5	cW6
1	1	82	158	229	298	0	0.2	0.2	0.2	0.2	0.2	0
2	1	0	0	0	0	0	1	0	0	0	0	0
3	1	15	136	153	174	294	0.167	0.167	0.167	0.167	0.167	0.167
4	1	162	218	362	0	0	0.25	0.25	0.25	0.25	0	0
5	1	14	0	0	0	0	0.5	0.5	0	0	0	0
6	1	0	0	0	0	0	1	0	0	0	0	0
7	1	0	0	0	0	0	1	0	0	0	0	0
8	1	0	0	0	0	0	1	0	0	0	0	0
9	1	0	0	0	0	0	1	0	0	0	0	0
10	1	176	0	0	0	0	0.5	0.5	0	0	0	0

column "cD6") for their health needs. There are other three patients (PID of #4, #5, and #10), who are also nested within multiple doctors (meaning seeing multiple doctors). For these three patients, the patient #4 visited four doctors (i.e., doctors #1, #162, #218, and #362), the patient #5 visited two doctors (i.e., doctors #1 and #14), and the patient #10 visited two doctors (i.e., doctors #1 and #176), respectively. The rest of the patients (PID #2, #6, #7, #8, and #9) are still seeing one doctor (in column "cD1"). The right panel of Table 13.1 shows the associated equal weighting (cW) as given to these patients across the doctors with equal weight. Thus, this assumes that each patient sees their doctors an equal numbers of time. In the absence of data to the contrary, this is the best one can do. These weights are easily adjusted as the researcher obtains evidence.

Figure 13.2 graphically illustrates the generated data nesting structure for the first 400 patients. The dots denote the patients (corresponding PID on the y-axis) who are assigned to the doctors (DID at the x-axis). At the left side of the figure, the solid-dotted line indicates the assignment prior to our adding multiple membership.

The above data generation process will add 407 extra columns to the original data for each of these 8,525 patients on which doctors they will visit among the 407 doctors as well as another 407 extra columns to indicate the corresponding weights on the proportion of the time they spend with each of these 407 doctors. Notice that the maximum number of doctors for each patient in this new dataset is 6; therefore, there are large proportion of the data that will be zeros.

Along with the newly generated 814 (407+407) columns, for each patient, the original data will be still kept for data analysis, and these variables are tumor size in millimeters (mm), level 1 predictors in the patients' age (Age), family history (FamilyHx), length of hospital stay, smoking history (SmokingHx), and cancer stage (CancerStage).

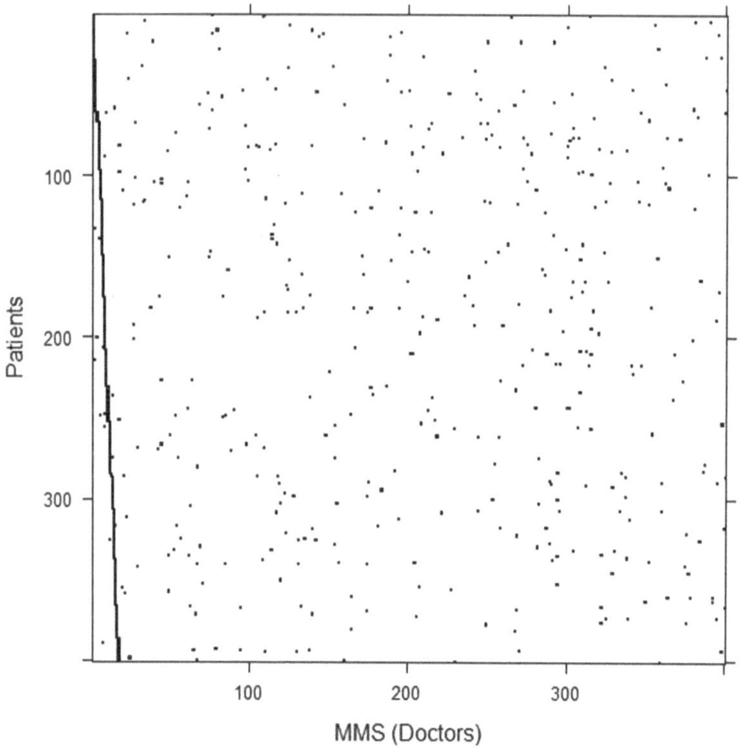

FIGURE 13.2
The Generated Hierarchical Data Structure for the First 400 Patients and 400 Doctors.

13.3 Two-Level Linear Mixed-Effects Models for Multiple Membership Data

There is an ongoing research on the development and analysis of multiple membership data. Some notable references are Browne et al. (2001), Fielding and Goldstein (2006), Arreola and Wilson (2017), and Arreola et al. (2020). An online search provides additional resources based on one's area of analysis. There are limited software development ready for use from the best of our knowledge. In this chapter, a description of multiple membership model with random intercept is outlined with the use of statistical software packages in SAS and R.

Corresponding to the HDP data considering the patients (Level 1) nested within doctors (Level 2), the two-level hierarchical multiple membership random intercept model can be specified as follows:

The level 1 model is as follows:

$$\text{Level 1: } y_{ij} = \beta_{0j} + \beta_{1j}x_{ij1} + \beta_{2j}x_{ij2} + \beta_{3j}x_{ij3} + \beta_{4j}x_{ij4} + \beta_{5j}x_{ij5} + \beta_{6j}x_{ij6} + \epsilon_{ij}$$

where i (i = 1, …, 8,525) denotes the ith patient, who is nested with jth (j = 1,…, 407) doctor; y_{ij} is the *tumor size*, x_{ij1} is *length of stay*, x_{ij2} is *age*, x_{ij3} is *FamilyHx*, x_{ij4} is *Smoking Hx*, x_{ij5} is *cancer stage*, and x_{ij6} is an interaction term of *SmokingHx* by *Age*. The error term ϵ_{ij} is assumed to be i.i.d. distributed with a normal distribution and within-cluster (doctor) variance σ^2, $\epsilon_{ij} \sim N(0, \sigma^2)$.

To model the between-cluster (doctor) variation in random intercept LME, we assume that the intercept parameter β_{0j} at level 2 varies, while the slope parameters are assumed the same across all clusters. Thus, the level 2 model can be specified as follows:

$$\text{Level 2: } \beta_{0j} = \beta_0 + \sum_{j \subset Doctor(i)} w_{ij}^{(2)} u_{0j}^{(2)}$$

where i = 1, …, N (N = 8,525) and j = 1,…, J (J = 407). The notation $j \subset Doctor(i)$ denotes the doctor j who belongs to the set of doctors the ith patient saw for his/her healthcare needs. The $w_{ij}^{(2)}$ denotes the weighting structure at level 2 for ith patient nested in jth doctor. The $u_{0j}^{(2)}$ denotes the level 2 random effects for jth doctor, which is modeled as i.i.d., random normally distributed with between-cluster (doctor) variance τ_0^2, i.e., $u_{0j}^{(2)} \sim N(0, \tau_0^2)$.

The generated multiple membership HDP data is used to illustrate the implementation of SAS and R for the two-level hierarchical multiple membership data. The HDP data are structured with patients (level 1) nested within doctors (level 2), and some of the patients have multiple membership multilevel structure to see multiple doctors for their health needs.

13.4 Data Analysis Using SAS

13.4.1 Data Format

Data are formatted to comply with SAS program requirement. The first step is to determine the maximum number of doctors the patients can see (i.e., six in the data HDP data). The second step is to create six columns to hold doctors. Each patient is nested for health care (Table 13.1). The third step is to create the associated weights as seen in Table 13.1. The complete dataset is created along with other variables, saved as an excel file named *"nhdpSAS.xlsx"*, and loaded into SAS.

The SAS program to load the data into SAS is as follows:

```
proc import datafile="Your Data Path\nhdpSAS.xlsx"
out=datnhdp
dbms=xlsx
replace; run;
```

The use of "proc import" to import the excel file into SAS, which is saved in SAS data file, is named as "datnhdp".

13.4.2 SAS Program

Having loaded SAS data, one can make use of SAS procedure "GLIMMIX" (i.e. generalized linear mixed-effects model) to fit the multiple membership model as follows:

```
proc glimmix data=datnhdp method=rspl;
class cD1-cD6 FamilyHx(ref="no")
          SmokingHx(ref="current") CancerStage(ref="I");
effect docs=multimember(cD1-cD6/weight=(cW1-cW6)
details);
model tumorsize = Age FamilyHx LengthofStay SmokingHx
          CancerStage SmokingHx*Age/solution cl;
random docs/solution;
run;
```

The procedure "proc glimmix" with "class" statement listing the 6 extra variables "cD1-cD6" for the multiple membership nesting structure where the patients are nested to a maximum of 6 doctors. The difference between single membership and multiple membership is in the code "effect docs=multimember (cD1-cD6/weight= (cW1-cW6) with details)". SAS "PROC GLIMMIX" that the doctors (i.e., "docs") have multiple membership nesting structure in the dataset with "cD1-cD6" along with the weights in "cW1-cW6". The random statement is for the doctors in which the patients are nested

The above SAS program will generate the following SAS output for multiple membership multilevel model fitting:

```
                    The GLIMMIX Procedure

                      Model Information

Data Set                          WORK.DATNHDP
Response Variable                 tumorsize
Response Distribution             Gaussian
Link Function                     Identity
Variance Function                 Default
Variance Matrix                   Not blocked
Estimation Technique              Restricted Maximum
                                  Likelihood
Degrees of Freedom Method         Kenward-Roger
Fixed Effects SE Adjustment       Kenward-Roger
```

Since the response is continuous the choice is to use an identity link and the estimation technique for the joint distribution used to obtain parameter estimates is restricted maximum likelihood. But other methods can be asked to conduct the estimation.

Selected Class Level Information

Class	Levels	Values
cD1	406	1 2 3 4 5 6 7 8 9 10 11 12 13 14 15 16 17 18 19 20 21 22 23 24 25 26 27
cD2	403	0 4 5 6 7 8 9 11 12 14 15 16 17 18 19 20 21 22 23 24 25 26 27 28 29 30
cD3	364	0 14 17 28 33 34 35 36 38 39 41 43 44 45 46 48 50 52 54 55 58 59 60 61
cD4	294	0 44 56 64 66 68 73 77 81 86 87 88 90 95 97 103 106 107 108 109 110
cD5	216	0 128 131 136 137 140 141 146 153 158 159 161 162 164 165 166 171

Selected Class Level Information

Class	Levels	Values
cD6	106	0 203 206 211 213 218 238 242 247 249 254 257 265 268 269 271 272
FamilyHx		
SmokingHx	3	former never current
CancerStage	4	II III IV I
Dimensions		
G-side Cov. Parameters		1
R-side Cov. Parameters		1
Columns in X		15

Convergence criterion (GCONV=1E-8) satisfied.

The G side parameter is on account of the random effects in doctors. The R side parameter is the usual error term

Covariance Parameter Estimates

Cov Parm	Estimate	Standard Error
docs	29.5594	2.4838
Residual	72.9242	1.1441

An estimate of the random effects variance is 29.56 with e.49 standard error. This suggest that doctors as a random effect was an important part of the model. Also the variance among the observation is large with a value of 72.92 and standard error of 1.14.

Effect	FamilyHx	Estimate	Standard Error	Pr>|t|	Lower	Upper
Intercept		69.8127	1.7810	<.0001	66.3215	73.3040
Age		0.1807	0.03655	<.0001	0.1091	0.2524
FamilyHx	yes	2.4313	0.2724	<.0001	1.8972	2.9654
FamilyHx	no	0	.	.	.	.
LengthofStay		0.7076	0.1103	<.0001	0.4914	0.9237
SmokingHx	former	-1.4464	2.4870	0.5609	-6.3215	3.4287

Effect	FamilyHx	Estimate	Standard Error	Pr>|t|	Lower	Upper
SmokingHx	never	-4.0154	2.0515	0.0504	-8.0369	0.006140
SmokingHx	current	0	.	.	.	.
CancerStage		1.9879	0.2556	<.0001	1.4869	2.4889
CancerStage		3.6607	0.3294	<.0001	3.0151	4.3064
CancerStage		6.2449	0.4282	<.0001	5.4054	7.0843
CancerStage		0	.	.	.	.
Age*SmokingHx	former	-0.2244	0.05028	<.0001	-0.3229	-0.1258
Age*SmokingHx	never	-0.3132	0.04143	<.0001	-0.3944	-0.2320
Age*SmokingHx	current	0	.	.	.	.

The fixed effects part of the model tells that age, family history, are significant in tumor size. There is a strong interaction between age and smoking history. The age of former and never smoker had a negative effect on the tumor size.

Selected subset of solution of Random effects

Effect	docs	Estimate	Std Err Pred	Pr > |t|
docs	1	-8.0356	1.8032	<.0001
docs	113	10.6131	2.1101	<.0001
docs	119	-11.4844	2.1207	<.0001
docs	127	5.5363	1.6375	0.0007
docs	128	-3.0135	2.1434	0.1598
docs	129	-8.8930	1.7455	<.0001
docs	13	-8.4206	2.1909	0.0001
docs	130	-2.9673	1.7464	0.0894

A subset of the doctors is included. The complete version is available online. Having care from doc#127 resulted in the patent having a larger tumor size. Whereas doc #119 resulted in the patient having smaller tumor size. This is by no means a cause and effect nor does it tell of the care obtained from a doctor. It simply shows how the interpretation of these estimates. The analyst needs more information before one can make any conclusions.

13.5 Data Analysis Using R

13.5.1 Data Format

The modeling of multiple membership and multilevel data using R package "lme" (i.e., linear mixed-effects model) and the data format are different from the data structure in SAS. The HDP data is used to illustrate the data format as per the requirement in R.

The 407 doctors in these data create an indicator (denoted here by "*W.csv*") to specify that patients are nested within doctors. Therefore, 407 columns are created for these 407 doctors. Within each of these 407 columns, each patient was mapped to the doctors. As seen in Table 13.1, the first patient (PID #1) visited five doctors, which are numbered as #1, #82, #158, #229, and #298. For this patient, we place 1 at columns 1, 82, 158, 229, and 298 and the rest of the columns as zeros. For a further example, for patient #2, put "1" for the first doctor and then 0 for the rest of the doctors since this patient only visited doctor #1. This indicator data is saved in "*W.csv*" to be loaded along with HDP data into R for multiple membership analysis.

13.5.2 Load the R Library and the Data

To model this HDP data using R, we make use of the R package "lme4"(i.e., linear mixed-effects) as follows:

```
# Call the R library
Library(lme4)
```

The data were saved in the data file "nhdpR.csv". We can load the data into R using the following R code:

```
# Load the Data
>nhdp = read.csv("nhdpR.csv", header=T)
# Other Data Characteristics to be Used For modeling
> ID.patient = sort(unique(nhdp$PID))
> n.patient = length(ID.patient);n.patient # 8525
> ID.doctor = sort(unique(nhdp$DID));ID.doctor
> n.doctor = length(ID.doctor);n.doctor #407
```

13.5.3 Random Intercept LME Model

Including all covariates in the LME model, we fit the random intercept LME model with the R code chunk as follows:

```
# Create a name list for doctors
> mm.names = as.factor(1:n.doctor);mm.names
> nhdp$mms = rep(mm.names,length.out=n.patient)
# Formulate the random-intercept LME model
> mms.int = lFormula(tumorsize ~ Age + FamilyHx +
      LengthofStay + SmokingHx + CancerStage +
      SmokingHx:Age + (1|mms),
data = nhdp)
# Replace the Z matrix
>mms.int$reTrms$Zt = mms.
int$reTrms$Ztlist[[1]] <- Matrix(t(W))
# Call Deviance function
>devfun = do.call(mkLmerDevfun, mms.int)
# Optimization
> opt = optimizeLmer(devfun)
> mms.int = mkMerMod(environment(devfun),
            opt, mms.int$reTrms,
fr = mms.int$fr)
# Print the summary  of model fitting
>summary(mms.int)
```

With the above R code, the output of the model fitting can be presented as follows:

```
Linear mixed model fit by REML ['lmerMod']
REML criterion at convergence: 61566.1

Random effects:
Groups  Name        Variance  Std.Dev.
mms     (Intercept) 28.54     5.343
Residual            73.62     8.580
Number of obs: 8525, groups:  mms, 407
```

An estimate of the random effects variance is 28.54 with standard error 5.343. This suggest that doctors as a random effect was an important part of the model. Also the variance among the observation is large with a value of 73.62 and stand error of 8.58.

Fixed effects:			
	Estimate	**Std.Error**	**t value**
(Intercept)	70.95636	1.79062	39.627
Age	0.16372	0.03677	4.453
FamilyHxyes	2.46027	0.27391	8.982
LengthofStay	0.65130	0.11106	5.865
SmokingHxformer	-2.52391	2.49511	-1.012
SmokingHxnever	-6.05856	2.06225	-2.938
CancerStageII	1.91840	0.25727	7.457
CancerStageIII	3.58189	0.33062	10.834
CancerStageIV	6.18567	0.42972	14.395
Age:SmokingHxformer	-0.20228	0.05043	-4.011
Age:SmokingHxnever	-0.27062	0.04162	-6.503

The output shows the random intercept LME model. The estimated within-doctor standard deviation is $\hat{\sigma} = 8.580$. The estimated between-doctor standard deviation for random intercept is $\hat{\tau}_0 = 5.343$.

The estimated fixed effects for "SmokingHxformer" is −2.52391 in reference to the "Current" smoking status and a standard error of 2.49511. This gives a *t*-value of −1.012, which is not statistically significant anymore. This conclusion is different from the results obtained in Chapter 5 where the smoking status is statistically significantly related to the "tumorsize". This difference is partially due to the multiple membership nested data structure, where larger variation existed within a multiple membership nesting data structure.

The fixed effects part of the model tells that age, family history, are significant in tumor size. There is a strong interaction between age and smoking history. The age of former and never smoker had a negative effect on the tumor size... The output shows for the random-intercept LME model, the estimated within-doctor standard deviation is $\hat{\sigma} = 8.580$. The estimated between-doctor standard deviation for random-intercept is $\hat{\tau}_0 = 5.343$. The estimated fixed-effects shows for

"SmokingHxformer" is -2.52391 in reference to the "Current" smoking status and a standard error of 2.49511. This gives a t-value of -1.012, which is not statistically significant anymore.

The information obtained using SAS and R produced similar results. This conclusion is different from the results obtained in Chapter 5 where the smoking status is statistically significantly related to the "tumorsize". That model was void of the opportunity to include in the model interaction as well as addressing the hierarchical data structure due to multiple memberships of patients. A larger variation is acknowledged due to the nesting data structure.

13.6 Discussions and Comments

In a random intercept multiple membership model, the intercepts are assumed random across clusters. In particular, in the HDP data, the random intercept multiple membership multilevel model assumes that all slopes between "tumorsize" and the "*Length of stay*" are the same for all 407 doctors regardless of their skills and other related factors. This may be too restricted assumption. It may be evident that the effect of the doctor differs based on their experience, and as such, one may want the assumption to allow slopes to vary, resulting in both the intercept and the slopes being random. This will produce at

$$\text{Level 1: } y_{ij} = \beta_{0j} + \beta_{1j}x_{ij1} + \beta_{2j}x_{ij2} + \beta_{3j}x_{ij3} + \beta_{4j}x_{ij4} + \beta_{5j}x_{ij5} + \beta_{6j}x_{ij6} + \epsilon_{ij}$$

where i (i = 1, ..., 8,525) denotes the ith patient, who is nested with jth (j = 1,...,407) doctor; y_{ij} is the tumorsize, x_{ij1} is the length of stay, x_{ij2} is age, x_{ij3} is familyHx, x_{ij4} is smokingHx, x_{ij5} is cancerStage, and x_{ij6} is an interaction term of smokingHx by age. The error term ϵ_{ij} is assumed to be i.i.d. distributed with a normal distribution and within-cluster (doctor) variance σ^2, $\epsilon_{ij} \sim N(0, \sigma^2)$.

Theoretically, the random intercept and the random slope multiple membership model assumes that the intercept parameter β_{0j} and the slope parameter β_{1j} at level 2 are random while assuming other parameters

$(\beta_{2j}, \beta_{3j}, \beta_{4j}, \beta_{5j}, \beta_{6j})$ are held fixed across clusters. The other part of the model is

$$\text{Level 2: } \beta_{0j} = \beta_0 + \sum_{j \subset \text{Doctor(i)}} w_{ij}^{(2)} u_{0j}^{(2)}$$

$$\beta_{1j} = \beta_1 + \sum_{j \subset \text{Doctor}(i)} w_{ij}^{(2)} u_{1j}^{(2)}$$

where $i = 1,\ldots N$ ($N = 8{,}525$) and $j = 1,\ldots, J$ ($J = 407$). The notation $j \subset \text{Doctor}(i)$ denotes the doctor j who belongs to the set of doctors the ith patient saw for his/her healthcare needs. The $w_{ij}^{(2)}$ denotes the weighting structure at level 2 for ith patients nested in jth doctor. The $u_{0j}^{(2)}$ and $u_{1j}^{(2)}$ denote the level 2 random effects for intercepts and slopes for jth doctor, which is modeled as i.i.d., random bi-normally distributed as follows:

$$\begin{pmatrix} u_{0j}^{(2)} \\ u_{1j}^{(2)} \end{pmatrix} \sim N\left(\begin{pmatrix} 0 \\ 0 \end{pmatrix}, \begin{pmatrix} \tau_0^2, & \tau_0\tau_1\rho_{01} \\ \tau_0\tau_1\rho_{01}, & \tau_1^2 \end{pmatrix}\right)$$

The implementation of this model follows readily in R for the interested readers.

13.7 Exercises

Data for three consecutive semesters consisting of 24,551 semester grade point averages (GPAs) for 14,103 undergraduate students enrolled in a college at a large university from May 2014 to May 2015 were obtained. The covariates of interest include student residency classification, student academic level, student-athlete or not, use of Disability Resources or not, and the number of classes enrolled in the semester. Approximately 59% of students are in-state residents, 37% of students are out-of-state residents, and 4% of students are international students. Approximately 1% of the students in the data are athletes. Less than 3% of students uses disability resources and services (DRS).

There are 40 majors, 1,867 courses, and 2,802 instructors. Approximately 3% of instructors are teaching assistants. Approximately 50% of the students are enrolled in business majors, with the remaining student majors as follows: 10% in accountancy-related majors or finance-related majors, approximately 7%

TABLE 13.2

Descriptive Statistics for Data

Variable	Semester 1	Semester 2	Semester 3
Residency status (%)			
In-state	65.62	61.54	56.97
Out-of-state	27.42	33.94	40.73
International	6.95	4.52	2.3
Academic level (%)			
Freshman	2.78	20.52	20.99
Sophomore	23.02	18.58	19.8
Junior	32.4	25.07	25.67
Senior	41.8	35.82	33.53
Average number of classes (SD)	4.65 (1.33)	4.70 (1.32)	4.76 (1.33)
Athlete (%)	0.88	1.05	0.82
DRS (%)	2.53	2.57	2.35

in marketing-related majors, 7% in management programs, approximately 6% in supply chain management, 3% in computer information systems, 2% in economics, and less than 1% in agribusiness. Descriptive statistics for variables in model per semester are provided in Table 13.2.

It is rare to find a student who took more than one class from the same instructor in a semester. Thus, assume that each of the instructors contributed equally to a student's semester GPA, regardless of the number of classes taken. Therefore, if student i has n instructors at semester t, then the weight $w_{i,j}$ for student i and instructor j was $1/n$ for each instructor. Therefore, the weights are assigned equally to the instructors. For students who only had one instructor, the corresponding weight is set to one. For students who have more than one semester of data, at semester t, nonzero weights are given only to instructors who teach the students at that semester. Weights for all other instructors including those who teach students in previous or later semesters where equal to zero. Using these data:

1. Fit a random effect multimembership model.
2. Outline the fixed effects and determine the ones of impact.
3. Look at the random effects used in this model.
    a. What is a measure of the variance component?
    b. Is it significant?
    c. Do you think that it was useful to address the multiple membership aspect?
4. What would you say to the dean of the college regarding the instructors on GPA?
5. What would you say to the dean regarding the demographics on GPA?

References

Arreola, E. V., Irimata, K., and Wilson, J. R. (2020). Common errors of interpretation in biostatistics. *Biostatistics and Epidemiology*, 4(1), 238–246. doi:10.1080/24709360.2020.1790085

Arreola, E. V. and Wilson, J. R. (2017). Analyzing multiple membership hierarchical data using PROC GLIMMIX. www.lexjansen.com/wuss/2017/100_Final_Paper_PDF.pdf

Browne, W. J., Goldstein, H., and Rasbash, J. (2001). Multiple membership multiple classification (MMMC) models. *Statistical Modeling*, 1(2), 103–124.

Fielding, A. and Goldstein, H. (2006). Cross-classified and multiple membership structures in multilevel models: An introduction and review. https://dera.ioe.ac.uk/id/eprint/6469/1/RR791.pdf.

Wilson, J. R., Chen, D. G., and Peace, K. E. (2023). *Statistical Analytics for Health Data Science with SAS and R*. Boca Raton, FL: Chapman and Hall/CRC Biostatistics Series.

14

Statistical Modeling of Survival Data

14.1 Research Interest/Question

Binary logistic regression models for independent observations (Chapter 7) and correlated observations (Chapters 9–11) in Wilson et al. (2023) identified factors associated with the presence or the absence of an event. In this chapter, the interest lies in factors that affect time related to that event, which is commonly called time-to-event data or more generally called survival data [e.g., time-to-the disease or time-to-death, etc.]. Other examples of events are time-to-divorce, or time-to-develop a tumor, etc. The outcome of time is continuous; therefore, a logistic regression model is not appropriate. Neither is the assumption that the outcomes are normally distributed, as time is positive and usually resembles a nonsymmetric distribution. Other models are needed to analyze the time-to-event data. Such models are commonly called survival models in cases, such as clinical, epidemiologic, or reliability research.

In studying survival models, the most important data type that one must consider is whether the observations are censored. Censored observations pertain to time when subjects leave the study for reasons other than the occurrence of the event. Censoring is present for some information about a subject's event time, but the exact event time is unknown. The censoring must be independent of the aspects related to survival. The times to the event or the survival times are nonnegative, and consequently, are usually positively skewed. When the observations are censored, the survival times are indicated with a certain numerical value.

This chapter discusses life tables, Kaplan–Meier, and Cox proportional hazard regression to model the survival time, or the time until the event. Though parametric models are mentioned, the discussion and examples are focused on nonparametric models.

In this chapter, the reader is introduced to survival analysis techniques used to model time until a single event using tools made available through statistical packages of SAS and R. More extensive exploration of survival

DOI: 10.1201/9781003596370-3

models are found in Cox and Oakes (1984), Therneau and Grambsch (2000), Hosmer et al. (2008), and Allison (2010, 2014).

14.2 HIV/AIDS Data

Survival data can be collected at several different situations, such as the time until tumor recurrence, time until cardiovascular death after some treatment intervention, time until AIDS for HIV patients, or time until a divorce, to name a few.

The data used in this chapter involves time-to-death in young infants with serious bacterial infection (Margolis et al. 1999). A subset of the data is shown in Table 14.1.

The complete dataset consists of 4,552 observations with 97 variables. In this subset dataset, the covariates are "Birthweight" and "Age" in days. There are two factors: "sleep more" and "cry more". The measure of the response variable is "time" to death. The "Status" is a binary variable that identifies whether or not the observation is censored with "Status = 1" if censored.

14.3 Principles of Survival Data Analysis

In survival data analysis, there are several methods for analyzing time-to-an-event. This statistical topic is listed under different names depending on the discipline to which it is applied. It is referred to as survival analysis in biostatistics, failure time analysis in engineering, reliability theory in industrial

TABLE 14.1

Subset of the HIV/AIDS Data

ID	Birthweight	Age	Sleep More	Cry More	Status	time
A0001	2595.9	62	0	0	0	145.00
A0002	2225.6	25	0	1	0	145.00
A0003	2298.4	25	0	0	0	145.00
A0005	3500.0	50	0	0	0	145.00
A0010	1650.0	80	0	1	0	145.00
A0011	3116.4	41	0	0	0	145.00
A0012	3000.0	56	0	0	0	145.00
A0013	3000.0	47	0	1	0	145.00
A0014	3550.0	30	0	1	0	145.00
A0015	2700.0	74	0	1	0	145.00

engineering, and event-history analysis in sociology. The response variable in survival data is the time-to-event, which is continuous in nature. Yet the measure of the response variable does not always have an exact time of an event for all subjects due to the specific property of censoring in survival data.

14.3.1 Censored Versus Uncensored Observations

There are censored and uncensored observations. The censored observations do not provide full value for a patient but partial information for the patient before the censoring. The censoring of participants refers to the problems of incomplete observations of time due to factors not related to the study design. There are subjects who may have to quit the study from some unrelated event (e.g., a move-away or a death) or simply do not have an event before the end of the observation period. The censoring techniques allow researchers to analyze incomplete data with partial information due to delayed entry or withdrawal from the study.

Complete observational time data are hardly ever realized in survival studies. The complete observations have the time-to-the-event to the best of one's knowledge. Thus, the observation is uncensored. Typically, the time-to-the-event is positively skewed. Its skewness is not a problem, as the aim is to model the probability of surviving beyond a certain time. The main interest is in the tail of the distribution.

14.3.2 Right Censoring

One usually encounters right censoring. This means that:

1. The subject does not experience the event, but the study ends.
2. The subject is lost to follow-up during the study period.
3. The subject withdraws from the study.

Such censoring is common. Right censoring consists of two kinds:

1. Fixed: the time is set for the duration of the study, and the subject does not experience the event.
2. Random: the study time is set, but censored subjects do not have all the same censoring time.

In all cases, the assumption is that censoring is caused by something other than the event under investigation. If there are no censoring, the standard regression procedures used to model continuous responses would be acceptable once one addresses the nonnormality. In addition, because of skewness, the mean of the distribution of the time-to-event would not be informative, but rather some function of it or some other parameter of the distribution is modeled.

14.4 Functions of the Distribution

Survival models are based on a response variable "time", which is continuous. Knowing that the probability of an event at a single point of a continuous distribution is zero. We concentrate on the cumulative distribution. Let T be a continuous random variable denoting the survival time.

14.4.1 Survival Function

Let the probability that a subject will have a survival time greater than or equal to t be denoted as $S(t)$. Then, the survival function is $S(t) = P(T > t) = 1 - F(t)$ that a subject will survive past the time t, where S stands for survival, P stands for probability, and $F(t) = P(T < t)$ stands for the cumulative distribution function (cdf). The survival function gives the probability of surviving or being event-free beyond time t. Because $S(t)$is a probability, it is positive and ranges from 0 to 1. It is defined as $S(0) = 1$ and as t approaches ∞, $S(t)$ approaches 0, as illustrated in Figure 14.1.

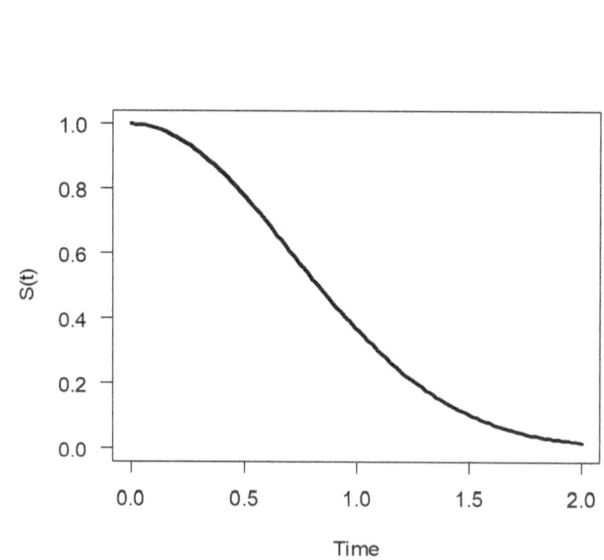

FIGURE 14.1
Illustration of Survival Function.

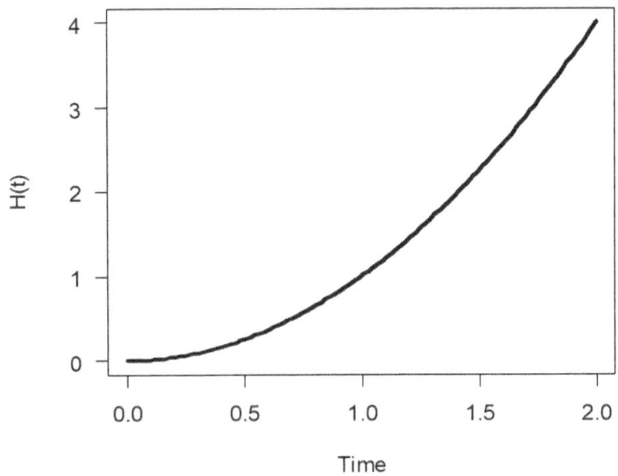

FIGURE 14.2
Cumulative Hazard Function.

14.4.2 The Hazard Function

The cumulative hazard function is defined as $H(t) = -\log(S(t))$. The instantaneous hazard function $h(t) = H'(t)$ is the instantaneous rate at which events occur, given no previous events. The hazard function describes the concept of the risk of an outcome (e.g., death, failure, hospitalization) in a small interval time $[t, t + s]$ (for small s) given that the individual subject has survived to time t.

Hazard and survival functions are mathematically related. If one models the hazard, then the survival is known and vice versa. The hazard, or hazard rate, is *not* a conditional probability. The hazard is never negative. It can be larger than 1. The hazard is the expected number of events per individual per unit of time as illustrated in Figure 14.2.

14.5 Survival Models

Although survival models are primarily used in the analysis of medical and biological sciences, it can also be used in the social sciences, econometrics, and engineering. It often begins with researchers wanting to know if there are differences in survival curves, and more so, how they differ graphically for different subpopulations. This analysis provides the estimates of survival function and certain plots. There are parametric and nonparametric models.

14.5.1 Parametric Models

A parametric model relies on a certain distribution of the outcome, i.e., time. Some distributions include as follows:

i. Exponential
ii. Weibull
iii. Gamma
iv. Lognormal

When a parametric model is used, the fit relies on the maximum likelihood method to estimate the unknown parameters of the distribution. A parametric linear regression model for the log hazard:

$$\log h(t) = \alpha + \beta_1 x_1 + \beta_2 x_2 + \ldots + \beta_p x_p$$

where x_i represents the ith covariate and β_i is the corresponding coefficient. The model is referred to as parametric if $\alpha, \beta_1, \ldots, \beta_p$ are specified, and $h(t)$ is fully specified based on the distribution. When x_i's are zero then $\log h(t) = \alpha$ representing the baseline hazard. However, it is not uncommon to have no knowledge of the distribution of $h(t)$. In this chapter, we do not discuss parametric models but rather concentrate more on nonparametric models.

14.5.2 The Kaplan–Meier Estimator: Nonparametric Model

The *Kaplan–Meier estimator* is the most commonly used method of estimating the survival function $S(t) = P(T > t)$. It can examine the relationship between survival and one or more covariates. The Kaplan–Meier estimator, or product limit estimator, is programmed in most statistical software packages as the first survival model. As the simplest procedure, it incorporates information from both censored and uncensored observations. It provides the survival curve that describes the relationship between the probability of survival and time, as illustrated in Figure 14.3.

Kaplan–Meier estimate is a simple but adequate method to measure the fraction of subjects surviving for a certain amount of time after treatment. In clinical research, the effect of a treatment or an intervention is assessed by measuring the number of subjects survived after that intervention over a period of time. The survival time can be affected by subjects under study who are disinterested and refused to remain in the study, or when some of the subjects may not experience the event or death before the end of the study; in other words, they are censored observations. The Kaplan–Meier estimate is one of the simplest ways of computing the survival over time in spite of

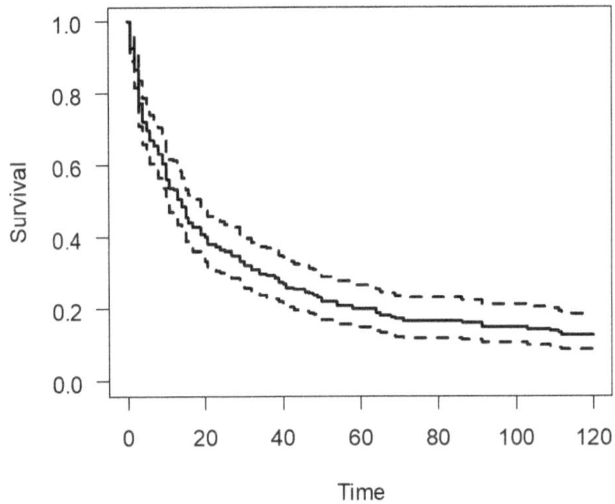

FIGURE 14.3
Illustration of Kaplan–Meier Graph.

censored observations. The survival curve is made assuming various situations. It computes probabilities of occurrence of event at a certain point of time and multiplying these successive probabilities by any earlier computed probabilities to get a final estimate. This can be calculated for multigroups of subjects, as well as their statistical difference in the survivals.

For each time interval, we compute the survival probability as the number of subjects surviving divided by the number at risk. We do not count those leaving for reasons other than the event, that is, subjects who are censored. Thus, the survival probability is

$$S_t = \frac{(\#\text{subjects living at the start}) - (\#\text{subjects died})}{(\#\text{subjects living at the start})}$$

The total probability of survival until that time interval is calculated by multiplying all the probabilities of survival at all time intervals before that time. Through the law of multiplication of probabilities, we calculate cumulative probability.

14.5.3 Cox's Proportional Hazards Regression Semiparametric Model

Another approach to modeling survival data uses a semiparametric method (Cox 1972). The approach does not specify the baseline hazard function $\alpha(t) = \log h_0(t)$ in

$$\log h(t) = \alpha + \beta_1 x_1 + \beta_2 x_2 + \ldots + \beta_p x_p$$

The baseline hazard can take any form, and the analysis is a proportional hazard model, meaning that it remains the same across time. The covariates in the systematic component through the linear predictor

$$\eta = \beta_1 x_1 + \beta_2 x_2 + \ldots + \beta_p x_p$$

has no constant term, as that is considered to be included in the baseline hazard.

The Cox regression model is well known as a proportional-hazards model (Allison 2010, 2014). Consider two subjects with different covariates, $\mathbf{x}_1$ and $\mathbf{x}_2$:

Subject 1

$$\mathbf{x}_1 = (x_{11}, x_{21}, \ldots, x_{p1})$$

$$\log h_1(\mathrm{t}) = h_0(t) + \beta_1 x_{11} + \beta_2 x_{21} + \ldots + \beta_p x_{p1}$$

Subject 2

$$\mathbf{x}_2 = (x_{12}, x_{22}, \ldots\ldots\ldots x_{p2}$$

$$\log h_2(t) = h_0(t) + \beta_1 x_{12} + \beta_2 x_{22} + \ldots + \beta_p x_{p2}$$

The hazard ratio for these two subjects is

$$\frac{h_1(\mathrm{t})}{h_2(\mathrm{t})} = \frac{\exp\left[h_0(t) + \beta_1 x_{11} + \beta_2 x_{21} + \ldots + \beta_p x_{p1}\right]}{\exp\left[h_0(t) + \beta_1 x_{12} + \beta_2 x_{22} + \ldots + \beta_p x_{p2}\right]}$$

$$= \exp\left[\left(\beta_1 (x_{11} - x_{12}) + \ldots + \beta_p \left(x_{p1} - x_{p2}\right)\right]\right.$$

This ratio is the same regardless of the time, since the $\mathbf{x}_1$ and $\mathbf{x}_2$ are free of time. The ratio, therefore, expresses the hazard for subject 1 relative to subject 2 at the time that the event occurred to the subject.

The Cox proportional hazards model is a particular form of hazards model. It predicts hazard as a combination of survival time and individual characteristics

$$h(t, \mathbf{x}, \beta) = h_0(t) . e^{x'\beta}$$

which is the (baseline hazard) × (individual effect: data value times regression coefficient β).

Cox's proportional hazards models look at the simultaneous effect of covariates on time-to-death, whereas in binary response data, the logistic regression model is used to model the odds as a measure of association. In the survival data, the survival model is used to model the risk as a measure of association. Cox models have no parametric assumptions. The estimates are obtained based on the use of a partial likelihood function from which a survival function is derived.

14.5.4 Interpretation of Regression Coefficient

Consider a model with one binary factor x_{i1} and one covariate, x_{i2}, then Cox's model is

$$\log h_i(t) = \alpha(t) + \beta_1 x_{i1} + \beta_2 x_{i2}$$

where $h_i(t)$ is the hazard function, $\alpha(t)$ is the baseline hazard, and β_1 and β_2 are regression coefficients. A simple interpretation of the measures of association in the Cox model as "relative risk" is very useful in modeling the risk of event for a certain covariate. When a two-level (dichotomous) covariate x_{i1} coefficient β_1 has a value of 0 = no and 1 = yes is observed, the hazard ratio becomes e^{β_1}. If the value of the coefficient is $\beta_1 = 0.783$, then the hazard ratio $e^{0.783} = 2.186$. So, with such a value, the subject labeled with a "1" (yes) is 2.186 times more likely to have an event than the subjects labeled with a "0" (no). Thus, we have a measure of association that gives insight into the difference between our exposure categories and those not being exposed (binary covariate).

In the model, when there is a continuous covariate x_{i2} with coefficient β_2 and a value 25 for a certain subject, as opposed to a value of 26 for another subject, then the hazard ratio becomes e^{β_2}. If the value of the coefficient is $\beta_2 = -0.01791$, then $e^{-0.01791} = 0.9822$. The measure says that the subject with a value of 25 instead of 26 or 45 instead of 46 (any two subjects with 1 unit apart), the risk decreases by 1 − 0.9822 = 0.0178. In this way, we have a measure of association that gives insight into the difference between the different measures on a continuous covariate.

14.5.5 Assumptions

The Cox proportional hazard model is an example of semiparametric modeling. It does not require assumptions to be made about the distribution of survival times. Thus, the method is considerably more robust. However, with this model, the researcher assumes that the hazards are proportional over time. Their ratio is assumed constant over the survival time. By proportional, we mean that if we compare a 25-year-old person with a 26-year-old person at time t, we do not see a difference in hazard ratio as if we were to check at another time s.

14.5.6 Estimation

The model parameters are estimated through a partial-likelihood estimation. This is not as efficient as the maximum likelihood estimation for the case of correctly specified distribution in the hazard regression model. Detailed explanations can be found from Cox and Oakes (1984) and Therneau and Grambsch (2000).

14.5.7 Cox Regression Diagnostics

The model relies on the assumption of constant hazards ratio. Therefore, it is important to know if this constant hazard ratio assumption is satisfied. If it is not, then a fitted Cox regression model does not adequately represent the data. Thus, it is crucial that the assumption of proportional hazards is a valid one. In addition, it is important that there is nonlinearity in the relationship between the log-hazard and covariates.

14.5.8 Extended Cox Proportional Hazard Model

It is often the case that the values of some covariates in a survival model change over time. Such covariates are referred to as time-dependent covariates, as they are explanatory variables that change over time as discussed in Lalonde et al. (2015). In survival analysis, the presence of time-dependent covariates warrant extra attention, as the ratios of the hazards will not remain steady, therefore, affecting the model structure. The Cox's proportional hazard model assumes that the ratio remains the same.

The extended Cox model has the flexibility to introduce time-dependent covariates and handle the censoring of survival times. This is done while using the partial likelihood function to obtain estimates. We present the extended Cox model with time-dependent covariates as

$$\log h(t) = \alpha(t) + \beta_1 x_1(t) + \beta_2 x_2(\mathrm{t}) + \ldots + \beta_p x_p(t)$$

In particular, if the researcher believes that there may be a time dependency from a certain variable, then simply add that variable by time-to-the-list of independent variables. If the interaction term is found to be insignificant, one can conclude that the proportional hazards assumption is satisfied. This is necessary to ensure that there is no adverse effect from time-dependent covariates that creates different rates for different subjects, thus making the ratios of their hazards not constant.

14.6 Data Analysis: Time-to-Death Study

This study investigates the effect of time-to-death while examining four covariates: "birthweight", "age in days", "sleep more", and "cry more". The indicator variable, "Status", distinguishes between censored and uncensored observation. The focus of the study is to see if "birthweight", "age" in days, "sleep more", and "cry more" had an impact on time-to-death. We use the statistical packages SAS and Rto fit the Kaplan–Meier and Cox's proportional hazards models.

14.6.1 Time-to-Death Study with SAS Programming

14.6.1.1 Kaplan–Meier Method

We first load the data into SAS with the following SAS code:

```
DATA SURVIVAL;
INPUT V1$ biwt age slpm hcm status time;
DATALINES;
A0001 2595.88330078125 62 0 0 0.00 145.00
A0002 2225.55224609375 25 0 1 0.00 145.00
A0003 2298.37207031250 25 0 0 0.00 145.00
A0005 3500.00000000000 50 0 0 0.00 145.00
A0010 1650.00000000000 80 0 1 0.00 145.00
A0011 3116.37353515625 41 0 0 0.00 145.00
.............................................................................
R0040 1572.77697753906 26 0 0 1.00  64.00
P0155 3180.00000000000  1 0 0 1.00 133.00;
```

With the data loaded, we can now fit the Kaplan–Meier survival curve using SAS procedure "lifetest" with the following SAS code:

```
proc lifetest data=SURVIVAL method=KM plots=S(NOCENSOR ATRISK CL);
time time*status(0);
run;
```

The above SAS code chunk would produce the following SAS output (Truncated Version of output):

The LIFETEST Procedure

Product-Limit Survival Estimates					
time	**Survival**	**Failure**	**Survival Standard Error**	**Number Failed**	**Number Left**
0.000	1.0000	0	0	0	4545
0.000	.	.	.	1	4544
0.000	.	.	.	2	4543
0.000	.	.	.	3	4542
0.000	.	.	.	4	4541
0.000	.	.	.	5	4540
145.000	*			1301	2
145.000	*			1301	1
145.000	*	0.7138		1301	0

The survival time and the survival rate along with its standard error are given at each stage. We truncated the table for brevity.

Summary Statistics for Time Variable time

Quartile Estimates				
		95% Confidence Interval		
Percent	**Point Estimate**	**Transform**	**[Lower**	**Upper)**
75	.	LOGLOG	.	.
50	.	LOGLOG	.	.
25	14.000	LOGLOG	12.000	17.000
Mean				Standard Error
97.219				0.841

Summary values for time are given. We truncated the table for brevity, and in particular, the mean of the time and standard error. Recall that survival distribution is skewed so the mean is not the best summary measure.

Summary of the Number of Censored and Uncensored Values			
Total	**Failed**	**Censored**	**Percent Censored**
4545	1301	3244	71.38

> There were 4545 subjects in the dataset. There are 71.38% of subjects censored.

14.6.1.2 Cox's Proportional Hazard Regression

The SAS procedure "phreg" is used to fit the Cox proportional hazard regression, and the SAS code is as follows:

```
proc phreg data=SURVIVAL;
model time*status(0)=age biwt slpm hcm/ ties=efron;
run;
```

> Status is the binary variable indicating censored. Ties =Efron is one of the SAS options. TIES=EFRON gives the researcher the approximations to the EXACT method without using the tremendous CPU it takes to run the EXACT method.

The PHREG Procedure

The associated SAS output from the above SAS program is as follows:
Model Information

Data Set	WORK.SURVIVAL
Dependent Variable	Time
Censoring Variable	Status
Censoring Value(s)	0
Ties Handling	EFRON

Number of Observations Read	4545
Number of Observations Used	4545

Summary of the Number of Event and Censored)Values

Total	Event	Censored	Percent Censored
4545	1301	3244	71.38

Convergence Status
Convergence criterion (GCONV=1E-8) satisfied.

Model Fit Statistics

Criterion	Without Covariates	With Covariates
-2 LOG L	21499.704	20812.699
AIC	21499.704	20820.699
SBC	21499.704	20841.382

Testing Global Null Hypothesis: BETA=0

Test	Chi-Square	DF	Pr > ChiSq
Likelihood Ratio	687.0053	4	<.0001
Score	769.2130	4	<.0001
Wald	735.6254	4	<.0001

The joint effect of the four covariates in the model is significant on time to event [p<.0001]. This is shown by any of the three tests [likelihood ratio, Score, and Wald].

Analysis of Maximum Likelihood Estimates

Parameter	DF	Parameter Estimate	Standard Error	Chi-Square	Pr > ChiSq	Hazard Ratio
age	1	-0.01791	0.00123	211.3869	<.0001	0.982
biwt	1	-0.0006952	0.0000493	199.1804	<.0001	0.999
slpm	1	0.79312	0.07620	108.3458	<.0001	2.210
hcm	1	0.57338	0.05600	104.8356	<.0001	1.774

Age, birthweight, sleep more, and cry more are significant predictors of time to death. Those who sleep more (slpm=1) and those cry more (hcm=1) have a greater risk for time to death than those who did not. Those with larger birthweight and those older in age had a smaller risk for time to death. Positive estimates say the hazard of death increases with for slpm=1 and hcm=1. While the negative estimates tell that those hazards decreases with age or birthweight.

14.6.2 Time-to-Death Study with R Programming

The analysis using R will be presented with detailed explanation following the R code as below.

```
## Load the "readxl" library to read the Excel data from "Your-Data-Path"
>library(readxl)
>dat = read_excel("survival.xlsx")
>attach(dat)
## Load the "survival" library to fit the Kaplan-Meier model
>library(survival)
>my.surv = Surv(time,status)
>Fit.KM = survfit(my.surv ~ 1)
## Print the model fitting
>summary(fit.KM)
## Call: survfit(formula = my.surv ~ 1)
##  time n.risk n.event survival std.err lower 95% CI upper 95% CI
## 0     4545      81      0.982 0.00196      0.978        0.986
## 1     4464      71      0.967 0.00267      0.961        0.972
## 2     4393      78      0.949 0.00325      0.943        0.956
## 3     4315      82       0.931 0.00375     0.924        0.939
## 4     4233      93      0.911 0.00423      0.903        0.919
## 5     4140     160      0.876 0.00489      0.866        0.885
## 6     3980      91      0.856 0.00521      0.846        0.866
## 7     3889     155      0.822 0.00568      0.811        0.833
## 8     3734      92      0.801 0.00592      0.790        0.813
## 9     3642      57      0.789 0.00605      0.777        0.801
## 10    3585      53      0.777 0.00617      0.765        0.789
## 57    3248       1      0.714 0.00670      0.701        0.728
## 64    3247       2      0.714 0.00670      0.701        0.727
## 133   3245       1      0.714 0.00670      0.701        0.727
```

The survival time and the survival rate along with its standard error are given at each stage. A 95% confidence interval for the survival rate is given at each point.

```
## Fit the Cox's Proportional Hazard Regression
>fit.cox = coxph(Surv(time,status)~ age+biwt+slpm+hcm, data=survival)
## Print the model fitting
>summary(res.cox)
## Call:
## coxph(formula = Surv(time, status) ~ age + biwt + slpm + hcm,
##    data = dat)
##  n= 4545, number of events= 1301
##     coef  exp(coef)  se(coef)    z Pr(>|z|)
```

```
## age    -1.791e-02  9.822e-01   1.232e-03 - 14.54  <2e-16 ***
## biwt   -6.952e-04  9.993e-01   4.926e-05 -14.11  <2e-16 ***
## slpm    7.931e-01  2.210e+00   7.620e-02  10.41  <2e-16 ***
## hcm     5.734e-01  1.774e+00   5.600e-02  10.24  <2e-16 ***
## ---
```

Age, birthweight, sleep more, and cry more are significant predictors of time to death. Those who sleep more (slpm=1) and those cry more (hcm=1) has a greater risk for time to death than those who did not. Those with larger birthweight and those older in age had a smaller risk for time to death. Positive estimates say the hazard of death increases, with for slpm=1 and hcm=1. The negative estimates tell that those hazards decrease with age or birthweight.

```
##        exp(coef) exp(-coef) lower .95 upper .95
## age      0.9822     1.0181    0.9799    0.9846
## biwt     0.9993     1.0007    0.9992    0.9994
## slpm     2.2103     0.4524    1.9037    2.5663
## hcm      1.7742     0.5636    1.5898    1.9801
```

Exp(coef) =hazard ratio. Subjects with slpm=1 have a higher hazard rate than those with slpm=0. Hazard ratios decrease with age and birthweight. It increases for slpm=1 and hcm=1.

```
## Concordance= 0.695  (se = 0.008 )
## Rsquare= 0.14  (max possible= 0.991 )
## Likelihood ratio test= 687    on 4 df,  p=0
## Wald test            = 735.6  on 4 df,  p=0
## Score (logrank) test = 769.2  on 4 df,  p=0
```

The overall effect of the four covariates are significant [likelihood ratio test, Wald test, and score test]. The joint effect is significant.

14.6.3 Results and Discussions

Four covariates (age, birthweight, sleep more and cry more) comprise the significant factors in the model. The adjusted hazard risk of 0.983 for age and 0.999 for birthweight means that as age or birthweight increases, the hazard decreases. Those who sleep more and those who cry more [hazard ratios of 2.188 and 1.767, respectively] are expected to have a larger hazard than those who do not. The robust nature of the Cox proportional hazard model allows one to approximate the results for the true parametric model when the true distribution is unknown or in question.

14.7 Exercises

The following variables are used from the dataset considered in this chapter: WBC, weight, heart rate, age in days, and hx stop breathing:

1. Fit Kaplan–Meier using the listed covariates
2. Fit Cox proportional hazard model
3. List the covariates in the hazard model that are significant
4. Test the proportion hazard ratio model assumption holds
5. Fit a Cox proportional hazards model and interpret the hazard ratios

References

Allison, P. (2010). *Survival Analysis Using the SAS System*, 2nd ed. Cary, NC: SAS Institute.

Allison, P. (2014). *Event History and Survival Analysis*, 2nd ed. Thousand Oaks, CA: Sage.

Cox, D.R. (1972). Regression models and life-tables. *Journal of the Royal Statistical Society. Series B (Methodological)*, *34* (2):187–220.

Cox, D. R. and Oakes, D. (1984). *Analysis of Survival Data*. London: Chapman and Hall.

Hosmer Jr., D., Lemeshow, S., and May, S. (2008). *Applied Survival Analysis*, 2nd ed. Hoboken, NJ: Wiley.

Lalonde, T., L., Wilson, J. R., and Yin, J. (2015). GMM logistic regression models for longitudinal data with time-dependent covariates and extended classifications. *Statistics in Medicine*, *33*(27), 4756–4769.

Margolis, P., Mulholland, E. K., Harrell, F., Gove, S., McCaul, F., Parker, S., Byass, P., Harrell, F., Mason, K., Carlin, J., Mulholland, K., Ogunlesi, O., Weber, M., Manary, M., Palmer, A., Adegbola, R., Whittle, H., Secka, O., Sam, B., ... Palmer,

P. (1999). Clinical prediction of serious bacterial infections in young infants in developing countries. *Pediatric Infectious Disease Journal*, *18*(10 SUPPL.), S23–S31. https://doi.org/10.1097/00006454-199910001-00005

Therneau, T. and Grambsch, P. (2000). *Modeling Survival Data*. New York, NY: Springer.

Wilson, J. R., Chen, D. G., and Peace, K. E. (2023). *Statistical Analytics for Health Data Science with SAS and R*. Chapman and Hall/CRC Biostatistics Series.

15

Statistical Modeling with Bayesian Paradigm

15.1 Research Interest/Question

In statistical modeling, there are two primary paradigms for statistical estimation and inference: frequentist and Bayesian. The frequentist paradigm treats probability as a limiting frequency, where the parameters in statistical models are fixed but estimated from a random sample, as discussed in previous chapters. In contrast, the Bayesian paradigm treats probability as a measure of the degree of belief in an unknown parameter, with the parameters in statistical models considered as random variables based on associated prior distributions. The Bayesian approach models the uncertainty of parameters using probabilities, allowing for sequential updates to these probabilities as more data become available.

This chapter provides an introductory overview of the Bayesian paradigm. For a more comprehensive discussion, readers are encouraged to refer to the works of Brooks and Roberts (1998), Gamerman and Lopes (2006), Carlin and Louis (2008), Brooks et al. (2011), Gelman et al. (2013), and Chen et al. (2017).

15.2 Public Health NHANES Data

In this chapter, we illustrate Bayesian modeling using the NHANES data from Chapter 5 and the Bangladesh data from Chapter 7, as presented in Wilson et al. (2023). We revisit these datasets within the Bayesian framework, comparing the results to those obtained from the frequentist paradigm in the multiple linear regression model from Chapter 5 and the logistic regression model in Chapter 7 of Wilson et al. (2023).

DOI: 10.1201/9781003596370-4

15.3 Bayes Theorem

Thomas Bayes (1702–1761) pioneered the Bayes theorem, a simple but elegant finding. The seminal Bayes' theorem is built on the concept of conditional probability. Consider the occurrence of two events (A and B) where the probability of event B is greater than zero, $P(B) > 0$. Define the conditional probability of A given B, $P(A \mid B)$, as the ratio of the joint probability of A and B, $P(AB)$), and the probability of B, $P(B)$ as $P(A \mid B) = \frac{P(AB)}{P(B)}$. Similarly, if $P(A) > 0$, then the conditional probability of B given A is $P(B \mid A) = \frac{P(AB)}{P(A)}$. From these two formulae of conditional probability, we can obtain that P(AB) = $P(B \mid A)P(A) = P(A \mid B)P(B)$, which leads to $P(A \mid B) = \frac{P(B \mid A)P(A)}{P(B)}$. This is the famous Bayes theorem. Thus, the Bayes theorem states that the conditional probability of event A given B, $P(A \mid B)$, is the product of its reversed conditional probability of event B given A, $P(B \mid A)$, and the event probability, $P(A)$, normalized by the probability of B, P(B).

Let us focus on the Bayes theorem according to a meaningful scenario, such as the sequential collection of data in social and health behavioral intervention research or sequential clinical trials. Assume that event A (for example, the treatment effect of clinical trials or any health intervention research from previous studies to be confirmed, which is why we call the probability of $P(A)$ as "prior" probability distribution) occurs before event B (for example, data collected from new clinical trials or intervention study). Now we conduct the new study and then observe the event B (i.e., the latest data). We can then calculate the conditional probability, $P(A \mid B)$, which is the probability of event A after event B occurred. This calculation is done based on Bayes theorem by the prior probability, $P(A)$, times the conditional probability, $P(B \mid A)$. This conditional probability, $P(A \mid B)$, is the "posterior probability" in Bayesian modeling.

15.4 Bayesian Modeling

In Chapter 3, we modeled the mean of a single population based on a frequentist hypothesis testing (Wilson et al. 2023). Our null hypothesis in terms of mean parameter value was $H_0: \theta = \theta_0$. We design studies and collect data (denoted by D) to test the hypothesis using the sample data, and the sampling distribution of the data given the parameter value (typically the null value of no effect), $P(D \mid H_0)$ or $P(D \mid \theta_0)$ can be evaluated based on how likely H_0 is true in order to declare to reject H_0.

15.4.1 Bayesian Modeling Principle

In the Bayesian paradigm, start with the prior distribution of the parameter $P(\theta)$ before data (D) is observed, thus representing the prior belief about how likely is the hypothetical parameter θ. Once the data (D) are collected, the data likelihood $P(D \mid \theta)$ is used, as in the frequentist paradigm. Using Bayes' theorem, we derive the posterior distribution of parameter, $P(\theta \mid D)$ as

$$P(\theta \mid D) = P(\theta)P(D \mid \theta)/P(D) \tag{15.1}$$

This posterior distribution is used to make statistical inferences about θ (i.e., the parameter of interest in the hypothesis) and the associated statistical inference. The conditional probability $P(\theta \mid D)$ is the posterior probability of θ after D is observed. $P(D \mid \theta)$ is the probability of observing D given θ, which is known as the likelihood in the frequentist paradigm, indicating the compatibility of the data with the hypothesis.

From Equation (15.1), the posterior probability of the parameter of interest is a combination of the prior belief of the likelihood of this parameter and the compatibility of the observed data in the data likelihood. Mathematically, the posterior is proportional to the prior times the likelihood. The foundation of Bayesian modeling lies in an elegant way that it combines new data with prior beliefs through the application of the Bayes theorem. It updates the parameter of interest iteratively and sequentially with the observed new data, the resulting posterior distribution is treated as a prior distribution, and a new posterior distribution emanates from new data. The Bayesian principle has the advantage in that it is applicable to various kinds of data, whether obtained all at once or over time.

This updating process is sequential, with the prior distribution being updated as more data are accumulated, leading to a new posterior distribution that refines statistical inferences. Thus, the Bayesian paradigm aligns with the general scientific principle of using cumulative information or knowledge to enhance inferences, as further discussed in Chen and Fraser (2017).

15.4.2 Bayesian Updating and Inference

Similar to frequentist inference, Bayesian inference involves both hypothesis testing and interval estimation. In Bayesian hypothesis testing, the posterior distribution is used to calculate the posterior probability that a particular hypothesis is true, given the observed data.

Bayesian interval estimates are derived from the posterior distribution and are often referred to as credible intervals. A credible interval is defined as an interval within which the parameter lies with a posterior probability of $1 - \alpha$.

For example, if the posterior probability that the parameter of interest, such as the treatment difference in a clinical trial, lies within an interval is 0.95, this interval is called a 95% *credible interval* ($\alpha = 0.05$).

The construction of credible intervals is straightforward in terms of how the interval is selected. Two commonly used types of credible intervals are the equal-tailed interval (with $1/2\alpha$ probability on each side) and the highest probability density (HPD) interval under the posterior distribution. For symmetric unimodal posterior distributions, these two intervals are identical. However, the HPD interval is often preferred in the Bayesian paradigm because it is the narrowest interval that covers the $1 - \alpha$ credible level, with all values within it being more probable than those outside of it.

The calculation of the HPD interval typically involves iterative methods to determine the narrowest interval that covers the $1 - \alpha$ probability and requires specialized computing algorithms. To perform these calculations, we rely on software such as R or SAS PROC MCMC.

15.4.3 Obtaining Information from Posterior: Markov-Chain Monte-Carlo

Due to the complexity of posterior distributions and the intractability of their analytical formulation, Markov-Chain Monte-Carlo (MCMC) methods are typically used in Bayesian inference. MCMC is a sampling technique that simulates data from posterior distributions, and it comprises two key components: "Markov-Chain" and "Monte-Carlo" simulation.

The "Markov-Chain" component refers to a sequence of random variables where the probability of moving to the next state depends only on the current state, not on the previous states, following a first-order autoregressive process. The "Monte-Carlo" component involves randomly simulating data from the posterior probability distribution. By taking a sufficient number of draws, the posterior distribution can be approximated with the desired level of precision.

When these two components are combined, the MCMC method constructs a valid Markov-Chain that converges to the target stationary posterior distribution. This allows for an accurate approximation of the desired posterior distribution, enabling statistical inferences to be made from the approximated distribution. In theory, the Markov-Chain should converge to the target posterior distribution, ensuring that the statistical inferences are valid. However, in practice, convergence can be challenging to achieve, and it is crucial to carefully examine convergence diagnostics to ensure the reliability of the results.

To construct a valid "Markov-Chain," methods such as the Gibbs sampler or the Metropolis–Hastings (MH) algorithm (Metreopilis, 1987) are commonly used. The MH algorithm is particularly versatile, as it utilizes a known and simpler distribution (e.g., a normal distribution) for simulation. This approach avoids the need for direct integration of the posterior distribution, making it especially attractive for Bayesian modeling.

15.4.4 Assessing Markov-Chain Convergence

Bayesian modeling requires the use of simulated draws to summarize the posterior distribution and calculate relevant quantities of interest. Researchers have cautioned that these simulation draws must be treated with care, as there are typically two major challenges. First, one must determine whether the Markov-Chain has reached its stationary, or desired posterior, distribution. Second, it is necessary to decide how many iterations to retain after the Markov-Chain has reached stationarity.

Convergence diagnostics are crucial for addressing these challenges. These diagnostics provide necessary—but not sufficient—conditions for confirming convergence. It is important to note that there are no definitive tests to conclusively determine when the Markov-Chain has fully converged to its stationary distribution.

Additionally, it is essential to check the convergence of all parameters, not just those of primary interest, before proceeding with any inference, as discussed in Carlin and Louis (2008). In some models, it is not uncommon for certain parameters to exhibit seemingly strong convergence behavior, which can be misleading if other parameters converge more slowly. If some parameters exhibit poor mixing, accurate posterior inference cannot be achieved for parameters that appear to have good mixing, as pointed out by Cowles and Carlin (1996) and Brooks and Roberts (1998).

15.4.5 Visual Analysis via Trace Plot

Trace plots, which display samples against the simulation index, are commonly used to assess convergence in Bayesian modeling. These plots help identify whether the Markov-Chain has converged to its stationary distribution or if a longer burn-in period is required. A well-constructed trace plot helps assess whether the chain mixes well. A Markov-Chain may have reached stationarity if the distribution of points in the trace plot remains consistent as the chain progresses. The most recognizable indicators of stationarity in a trace plot are a relatively constant mean and variance. A well-mixing chain rapidly traverses its posterior space, moving from one region of the posterior to another in relatively few steps.

15.5 Bayesian Data Analysis: Continuous Outcome

We revisit the data from Chapter 5, where frequentist multiple regression models for independent observations were analyzed. In that analysis, age, waist circumference, and albumin levels were used to model BMI. In this chapter, we introduce Bayesian multiple regression models using two statistical programs, SAS and R.

15.5.1 Data Analysis Using SAS Program

The Bayesian analysis can be implemented using the SAS procedure "GENMOD" with "Bayes" option as illustrated in the following SAS code chunk:

```
ods graphics ON;
proc genmod data=BMI;
model bmi = age waist albumin / dist=normal link=identity;
bayes seed=1 coeffprior=normal;
run;
```

> We use the GENMOD procedure with the Bayes option to fit Bayes multiple regression. The distribution (DIST) is normal since it is what we assumed in the multiple regression model. The identity link is used since the mean is modeled on the same scale as the data. The prior for coefficients (COEFFPRIOR) is normal.

The associated SAS output with above SAS program is as follows with detailed explanations in the text boxes:

THE GENMOD PROCEDURE: BAYESIAN ANALYSIS

Model Information	
Data Set	WORK.BMI
Burn-In Size	2000
MC Sample Size	10000
Thinning	1
Sampling Algorithm	Conjugate
Distribution	Normal
Link Function	Identity
Dependent Variable	bmi
Number of Observations Read	6490
Number of Observations Used	6490

Analysis Of Maximum Likelihood Parameter Estimates

Parameter	DF	Estimate	Standard Error	Wald 95% Confidence Limits	
Intercept	1	4.4655	0.3179	3.8423	5.0887
age	1	0.0071	0.0031	0.0010	0.0133
waist	1	0.2310	0.0031	0.2249	0.2372
albumin	1	0.3301	0.0078	0.3148	0.3454
Scale	1	5.0024	0.0439	4.9171	5.0892

Age, waist, and albumin are significant covariates in explaining the BMI. This is seen through the 95% confidence limits.

Independent Prior Distributions for Model Parameters	
Parameter	Prior Distribution
Dispersion	Improper

Initial Values of the Chain

Chain	Seed	Intercept	age	Waist	albumin	Dispersion
1	1	4.465499	0.007146	0.231031	0.33008	25.01667

Fit Statistics	
DIC (smaller is better)	39324.66
D (effective number of parameters)	5.003

Posterior Summaries

Parameter	N	Mean	Standard Deviation	Percentiles 25%	50%	75%
Intercept	10000	4.4616	0.3159	4.2487	4.4646	4.6735
age	10000	0.00714	0.00314	0.00498	0.00713	0.00924
waist	10000	0.2311	0.00314	0.2290	0.2310	0.2332
albumin	10000	0.3301	0.00782	0.3249	0.3302	0.3353
Dispersion	10000	25.0473	0.4428	24.7461	25.0454	25.3445

Posterior Intervals

Parameter	Alpha	Equal-Tail Interval		HPD Interval	
Intercept	0.050	3.8384	5.0775	3.8646	5.0995
age	0.050	0.000905	0.0134	0.00122	0.0136
waist	0.050	0.2249	0.2373	0.2251	0.2374
albumin	0.050	0.3145	0.3453	0.3149	0.3457
Dispersion	0.050	24.1888	25.9316	24.2228	25.9614

The summary of the posterior distribution for the regression coefficients and the variance are given. The percentiles are provided.

Posterior Correlation Matrix

Parameter	Intercept	age	Waist	albumin	Dispersion
Intercept	1.000	-0.147	-0.881	-0.542	-0.005
age	-0.147	1.000	-0.283	-0.203	-0.001
waist	-0.881	-0.283	1.000	0.522	0.006
albumin	-0.542	-0.203	0.522	1.000	-0.001
Dispersion	-0.005	-0.001	0.006	-0.001	1.000

The correlation among the regression estimates are given. The diagonal, as expected, has a value of one.

Posterior Autocorrelations

Parameter	Lag 1	Lag 5	Lag 10	Lag 50
Intercept	-0.0110	0.0093	-0.0075	-0.0037
age	-0.0204	-0.0114	-0.0091	0.0026
waist	-0.0177	-0.0052	-0.0125	0.0019
albumin	-0.0088	0.0134	-0.0173	0.0080
Dispersion	0.0041	-0.0091	-0.0030	-0.0000

Geweke Diagnostics

Parameter	z	Pr > |z|
Intercept	0.4946	0.6209
age	0.1570	0.8752
waist	-0.4450	0.6563
albumin	-0.4002	0.6890
Dispersion	1.1051	0.2691

Geweke's diagnostic is used to determine the burn-in period. (The smallest early portion of the chain that passes the diagnostic.) It mimics the simple two-sample test of means. If the mean of the first 10% is not significantly different from the last 50%, then we conclude the target distribution converged somewhere in the first 10% of the chain. The p-values [Pr > |z|] suggest convergence.

Effective Sample Sizes			
Parameter	**ESS**	**Autocorrelation Time**	**Efficiency**
Intercept	10224.6	0.9780	1.0225
age	10425.9	0.9591	1.0426
waist	10367.0	0.9646	1.0367
albumin	10000.0	1.0000	1.0000
Dispersion	10000.0	1.0000	1.0000

The MCMC convergence can be diagnosed in the following figures where a stabile and consistent MCMC sample chain indicates the estimates of MCMC converges to the target distribution for the intended parameter. This is shown in the following Figure 15.1 for the "Intercept", Figure 15.2 for the regression parameter of "Age", Figure 15.3 for the regression parameter of "Waist", Figure 15.4 for the regression parameter of "Albumin", and Figure 15.5 for the dispersion parameter "sigma".

15.5.2 Data Analysis Using R

There are many R packages for Bayesian modeling and data analysis. Historically, most of the R packages are developed to call *WinBUGS*, the

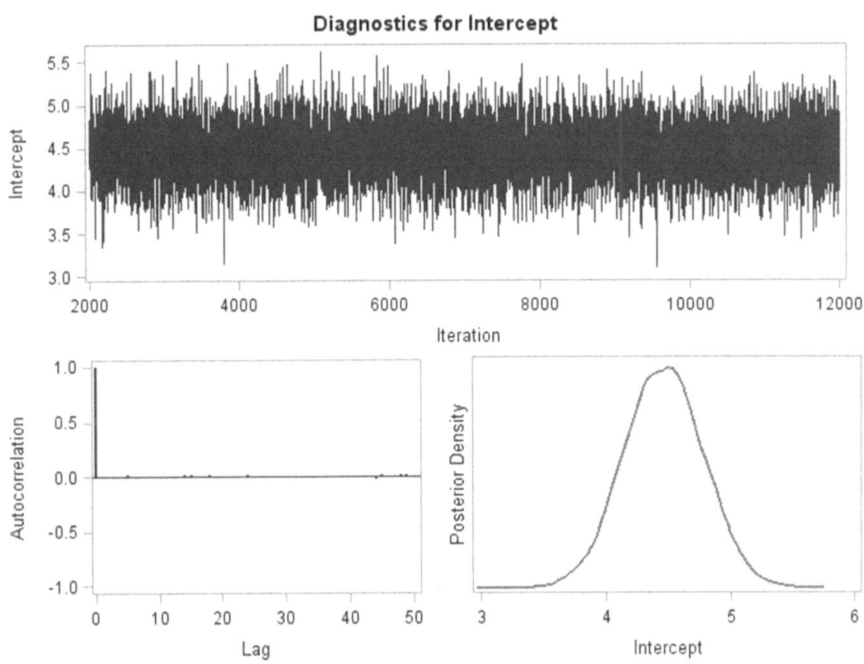

FIGURE 15.1
MCMC Estimation of the Intercept Parameter from SAS Proc MCMC.

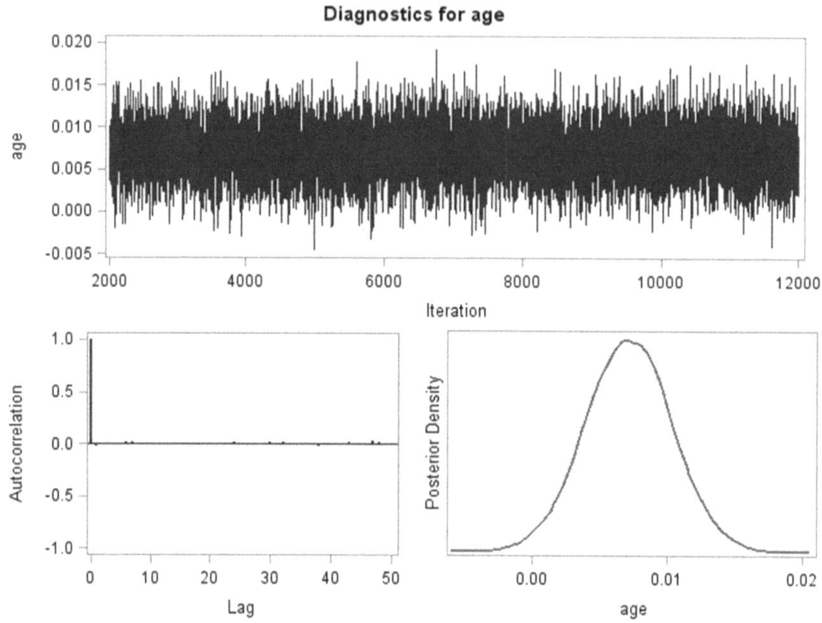

FIGURE 15.2
MCMC Estimation of the Regression Parameter for "Age" from SAS Proc MCMC.

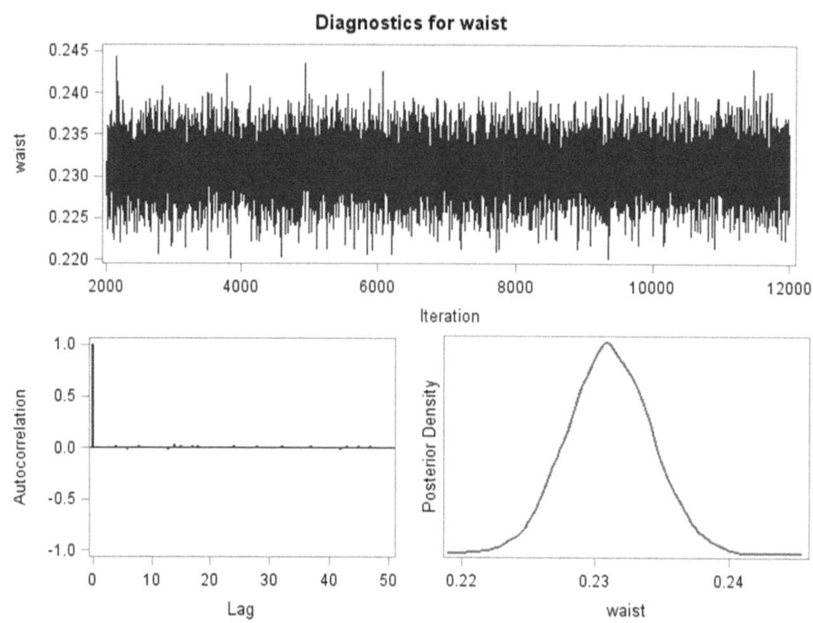

FIGURE 15.3
MCMC Estimation of the Regression Parameter for "Waist" from SAS Proc MCMC.

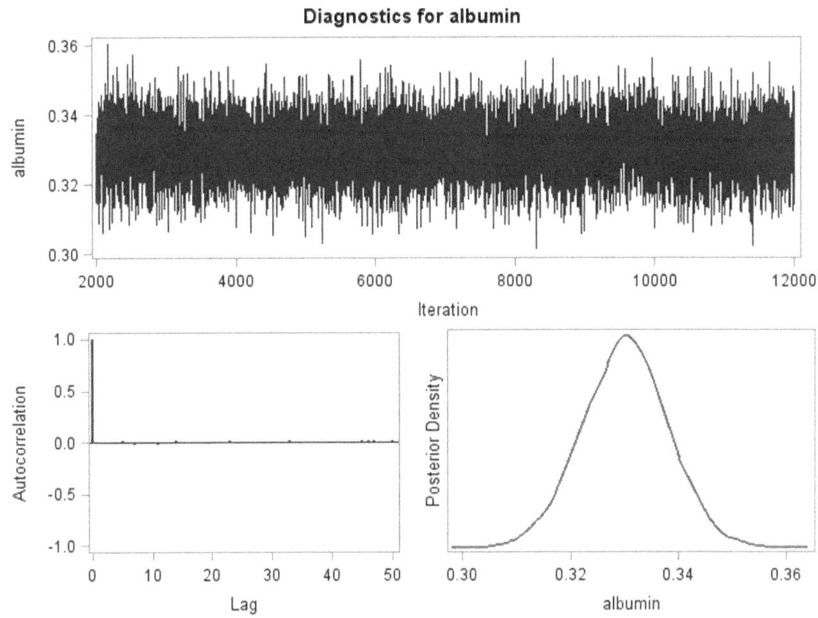

FIGURE 15.4
MCMC Estimation of the Regression Parameter for "Albumin" from SAS Proc MCMC.

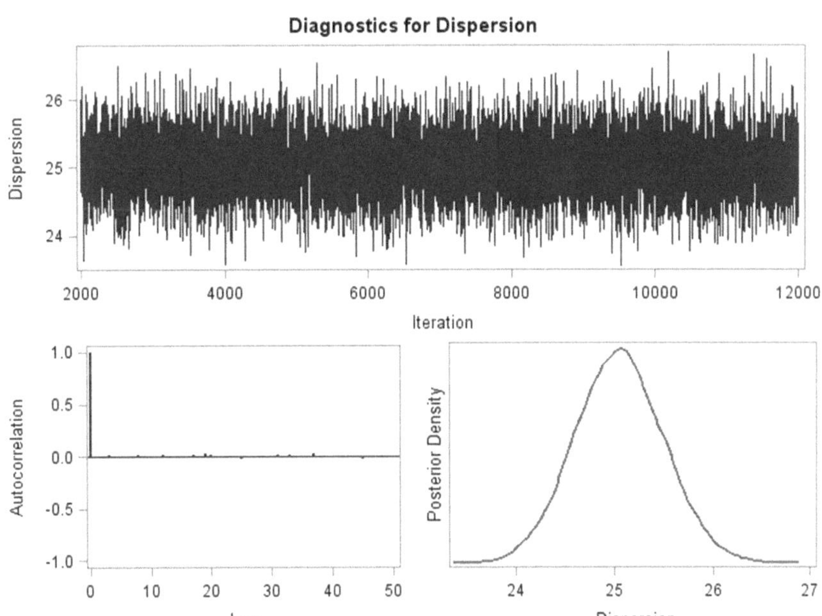

FIGURE 15.5
MCMC Estimation of the Dispersion Parameter "sigma" from SAS Proc MCMC.

Windows version of *BUGS* (*B*ayesian Inference *U*sing *G*ibbs *S*ampling), such as *R2WinBUGS*, *BRugs*, *rbugs*. WinBUGS incorporates Gibbs sampling and the Metropolis–Hastings algorithm to generate a Markov-Chain by sampling from full conditional distributions. It can be freely downloaded from www.mrc-bsu.cam.ac.uk/software/bugs-project. Typically, these packages call WinBUGS to follow its model building steps with three files that are needed to build a Bayesian model and run the program. These three files include a "model" file, a "data" file, and an "initial value" file. The model file is written by the user with syntax from WinBUGS, which is R code. Therefore, the user needs to be good at programming in both R and WinBUGS. The "data" and "initial" files are also built using R syntax. With these files, the packages will then use a function call to WinBUGS to run the program and output the results. The output can then be analyzed in R for convergence and relevant output analysis. The flexibility of these packages to use WinBUGS is an advantage for advanced and experienced users who want to build and fit advanced custom models.

Different from all these packages to call WinBUGS, a stand-alone R package *MCMCpack* was designed by Andrew D. Martin, Kevin M. Quinn, and Jong Hee Park. This package implements model-specific MCMC algorithms done by R function calls, which is very similar to other function calls in typical data analysis. Therefore, for unexperienced R users, it is easy to use. Details about this package are found at https://cran.r-project.org/web/packages/MCMCpack/index.html. We make use of this package in the chapter to implement the Bayesian data analysis. We present the Bayesian multiple regression using R package *MCMCpack*. The R program to implement the Bayesian analysis is as follows with detailed explanations.

```
> # load the R library "MCMCpack"
>library(MCMCpack)
> # for help and detailed description of this package, use "help"
>library(help=MCMCpack)
```

> This call provides a separate page to explain the extensive functionalities of this R package as described in the package description to contain functions to perform Bayesian inference using posterior simulation for a number of statistical models. Most simulation is done in compiled C++ written in the Scythe Statistical Library Version 1.0.3. All models return coda MCMC objects that can then be summarized using the coda package. Some useful utility functions such as density functions, pseudo-random number generators for statistical distributions, a general purpose Metropolis sampling algorithm, and tools for visualization are provided.

```
># read the data into R and name it as "nhgh"
>nhgh = read.csv("newnhgh.csv", header=T)
># call "lm" for frequentist multiple regression
>reg.nhgh = lm(BMI~age+waist+albumin, data=nhgh)
># print the summary of regression
> summary(reg.nhgh)
Call:
lm(formula = BMI ~ age + waist + albumin, data = nhgh)
Coefficients:
                  Estimate Std. Error  t value   Pr(>|t|)
(Intercept)   2.157392   0.541418     3.985    6.83e-05 ***
age          -0.054258   0.001684   -32.210   < 2e-16 ***
waist         0.375664   0.002056   182.758   < 2e-16 ***
albumin      -1.804125   0.104676   -17.235   < 2e-16 ***
Signif. codes:  0 '***' 0.001 '**' 0.01 '*' 0.05 '.' 0.1 ' ' 1
Residual standard error: 2.532 on 6471 degrees of freedom
(320 observations deleted due to missingness)
Multiple R-squared:  0.8596,   Adjusted R-squared:  0.8595
F-statistic: 1.32e+04 on 3 and 6471 DF, p-value: < 2.2e-16
```

This reproduced the results in Section 5.3. For Bayesian regression, we call the function "MCMCregress" implemented in the MCMCpack as follows:

```
> # call "MCMCregress" for Bayesian regression
> MCMCreg.nhgh = MCMCregress(BMI~age+waist+albumin, data=nhgh)
> # print the summary of Bayesian regression
> summary(MCMCreg.nhgh)
```

This produces a summary of this Bayesian data analysis as follows:

```
Iterations = 1001:11000
Thinning interval = 1
Number of chains = 1
Sample size per chain = 10000
```

```
1. Empirical mean and standard deviation for each variable,
   plus standard error of the mean:
                 Mean        SD      Naive SE  Time-series SE
(Intercept)   2.15172  0.540523  5.405e-03       5.405e-03
age          -0.05428  0.001707  1.707e-05       1.707e-05
waist         0.37568  0.002048  2.048e-05       1.997e-05
albumin      -1.80285  0.104206  1.042e-03       1.042e-03
sigma2        6.41036  0.112578  1.126e-03       1.141e-03
```

The Bayesian estimation is very close to the frequentist linear regression since a non-informative prior is used in this analysis.

```
2. Quantiles for each variable:
                 2.5%      25%      50%      75%    97.5%
(Intercept)   1.10622  1.79003  2.15645  2.50776  3.24638
age          -0.05764 -0.05542 -0.05429 -0.05313 -0.05092
waist         0.37161  0.37433  0.37569  0.37705  0.37965
albumin      -2.01094 -1.87276 -1.80334 -1.73228 -1.60204
sigma2        6.19424  6.33326  6.40845  6.48483  6.63731
```

In summary, the Bayesian modeling is implemented using the function "MCMCregress," which is similarly implemented in the frequentist multiple linear regression using "lm". It prints the summary information of the MCMC for various quantities, the posterior mean, standard deviation, and quantiles from the MCMC samples. The results from this Bayesian analysis are similar to the multiple regression in Section 5.3 because the noninformative prior was used. If prior information is available for inclusion, then options in "MCMCregress" are modified to include such information. At the beginning of the summary information, the MCMC ran 11,000 times, and with 1000 burn-in out samples (i.e., these samples are discarded due to the initial "Markov-Chain" correlation), resulting in 10,000 samples for the Bayesian analysis due to "Thinning interval" set to 1. These options are easily modified for more complicated Bayesian data, where the convergence might be problematic.

Similarly, the MCMC convergence from R package "MCMCregress" can be diagnosed in the following figures where a stabile and consistent MCMC sample chain indicates the estimates of MCMC converges to the target distribution for the intended parameter. As seen below, Figure 15.6 is for the "Intercept", regression parameter of "Age", and the regression parameter of "Waist", and Figure 15.7 for the regression parameter of "Albumin" and the dispersion parameter "sigma2".

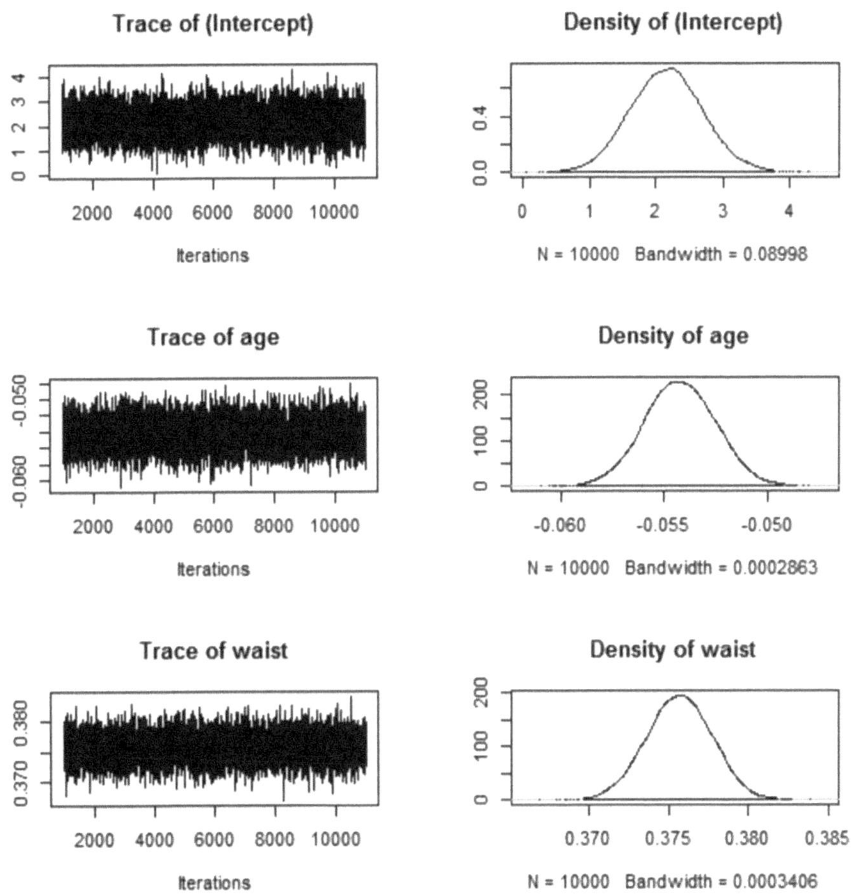

FIGURE 15.6
MCMC Estimation of the "Intercept" Parameter, Regression Parameter for "Age" and Regression Parameter for "Waist" from R Package "MCMCRegress".

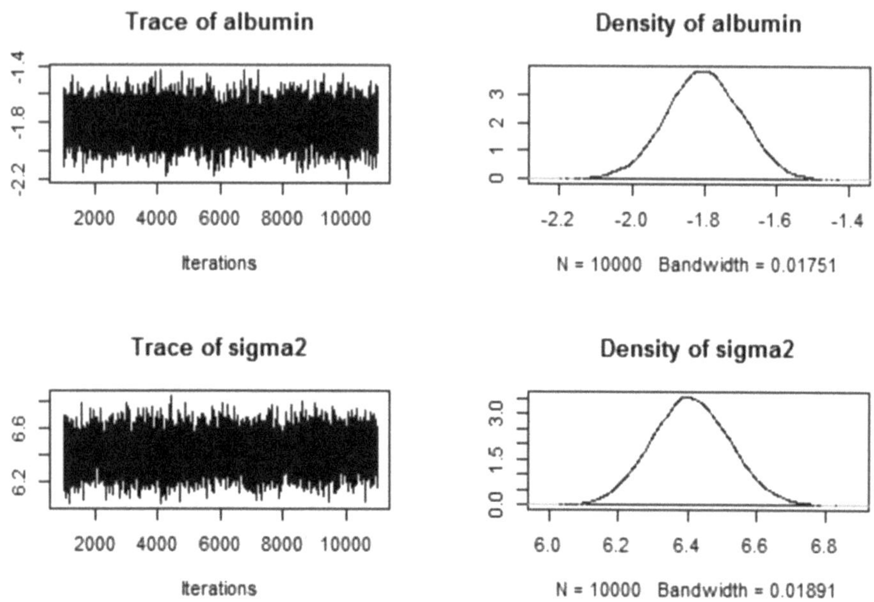

FIGURE 15.7
MCMC Estimation of the "Albumin" and the Dispersion Parameter of "sigma2" from R Package "MCMCRegress".

15.6 Bayesian Data Analysis: Binary Outcome

Bayesian modeling of binary data is demonstrated using the Bangladesh Demographic and Health Survey data, previously analyzed in Chapter 7 (Wilson et al. 2023) using a frequentist logistic regression on contraceptive use (CUC). The predictors in this analysis include age, education, television in the home, religious practice, and number of children. In this chapter, we apply a Bayesian logistic regression model using both SAS and R.

15.6.1 Data Analysis using SAS Programming for Binary Response

The SAS program to implement the Bayesian analysis using "Proc GENMOD" is as follows:

```
ods graphics ON;
proc genmod data=CUC;
class education;
model CUC=age children education religion TV/dist=binomial link=logit;
bayes seed=1 coeffprior=normal;
run;
```

We use the GENMOD procedure with the Bayes option to fit Bayes logistic regression. The binomial distribution (DIST) is assumed for logistic regression modeling of binary outcomes. The link is logit since we model the mean (proportion) on the log of the odds scale. The prior for coefficients (COEFFPRIOR) is normal. SAS access MCMC for logistic regression is provided through the Bayes statement in proc genmod. There are several default priors available. The normal prior is the most flexible (in the software), allowing different prior means and variances for the regression parameters.

The associated SAS output from the above SAS program is presented below with detailed explanations in the textboxes. All the MCMC samples for all the regression parameters are converged, and therefore, the MCMC convergence figures are not shown in this section.

The GENMOD Procedure
Bayesian Analysis
Model Information

Data Set	WORK.CUC
Burn-In Size	2000
MC Sample Size	10000
Thinning	1
Sampling Algorithm	Gamerman
Distribution	Binomial
Link Function	Logit
Dependent Variable	CUC
Number of Observations Read	16186
Number of Observations Used	16186
Number of Events	6508
Number of Trials	16186

Class Level Information

Class	Levels	Values
education	4	0 1 2 3

Response Profile

Ordered Value	CUC	Total Frequency
1	0	6508
2	1	9678

Analysis Of Maximum Likelihood Parameter Estimates

Parameter	**DF**	**Estimate**	**Standard Error**	**Wald 95% Confidence Limits**	
Intercept	1	-1.2334	0.1008	-1.4309	-1.0359
age	1	0.0482	0.0024	0.0435	0.0528

Analysis Of Maximum Likelihood Parameter Estimates						
children		1	-0.3171	0.0143	-0.3450	-0.2891
education	0	1	0.6729	0.0738	0.5282	0.8175
education	1	1	0.4908	0.0709	0.3518	0.6299
education	2	1	0.4909	0.0678	0.3580	0.6238
education	3	0	0.0000	0.0000	0.0000	0.0000
religion		1	-0.3303	0.0478	-0.4240	-0.2366
TV		1	-0.1350	0.0355	-0.2046	-0.0653
Scale		0	1.0000	0.0000	1.0000	1.0000

Age, children, education, religion, and television are significant covariates in contraceptive use. The significance is determined by the fact that confidence limits do not contain the value zero.

Bayesian Analysis

Posterior Summaries

Parameter	N	Mean	Standard Deviation	Percentiles 25%	50%	75%
Intercept	10000	-1.2329	0.1018	-1.3019	-1.2326	-1.1656
age	10000	0.0482	0.00240	0.0466	0.0482	0.0498
children	10000	-0.3172	0.0143	-0.3269	-0.3173	-0.3076
education0	10000	0.6725	0.0735	0.6216	0.6724	0.7227
education1	10000	0.4902	0.0718	0.4414	0.4898	0.5388
education2	10000	0.4908	0.0682	0.4445	0.4908	0.5370
religion	10000	-0.3312	0.0478	-0.3631	-0.3309	-0.2995
TV	10000	-0.1354	0.0355	-0.1594	-0.1356	-0.1111

The posterior distributions have mean, standard deviation, and percentiles listed. The posterior distribution tells about the significance of the covariates.

Posterior Intervals

Parameter	Alpha	Equal-Tail Interval		HPD Interval	
Intercept	0.050	-1.4305	-1.0306	-1.4365	-1.0406
age	0.050	0.0435	0.0529	0.0435	0.0528
children	0.050	-0.3450	-0.2891	-0.3451	-0.2893
education0	0.050	0.5283	0.8170	0.5326	0.8206
education1	0.050	0.3508	0.6316	0.3502	0.6305
education2	0.050	0.3585	0.6258	0.3573	0.6242
religion	0.050	-0.4246	-0.2374	-0.4237	-0.2371
TV	0.050	-0.2049	-0.0654	-0.2041	-0.0649

The posterior intervals [equal-tail interval & HPD interval] are given for the regression coefficients. The HPD region is an interval whe the posterior is unimodal. The Bayesian set estimates are called *credible sets,* which are also known as *credible intervals.* This is analogous to the concept of confidence intervals used in classical statistics. The HPD is an interval in which most of the distribution lies. Some statisticians prefer this interval because it is the smallest interval.

The GENMOD Procedure
Bayesian Analysis
Posterior Autocorrelations

Parameter	Lag 1	Lag 5	Lag 10	Lag 50
Intercept	0.1151	-0.0024	0.0020	-0.0144
age	0.1302	-0.0070	-0.0165	-0.0076
children	0.1241	0.0083	-0.0145	0.0111
education0	0.1180	-0.0088	-0.0010	-0.0040
education1	0.1176	0.0028	0.0013	-0.0048
education2	0.1292	0.0063	0.0106	-0.0005
religion	0.1064	0.0030	-0.0115	0.0069
TV	0.1244	0.0164	-0.0091	-0.0069

Geweke Diagnostics

Parameter	z	Pr > |z|
Intercept	0.0127	0.9899
age	0.1915	0.8481
children	0.3593	0.7194
education0	-0.4213	0.6735
education1	-0.5615	0.5745
education2	-0.3759	0.7070
religion	0.4871	0.6262
TV	-3.0133	0.0026

Geweke's diagnostic is used to determine the burn-in period. (The smallest early portion of the chain that passes the diagnostic.)It is similar to the two-sample test of mean. It mimics the simple two-sample test of means. If the mean of the first 10% is not significantly different from the last 50%, then we conclude the target distribution converged somewhere in the first 10% of the chain. The p-values [Pr > |z|] suggest convergence.

Effective Sample Sizes			
Parameter	**ESS**	**Autocorrelation Time**	**Efficiency**
Intercept	8129.2	1.2301	0.8129
age	7768.8	1.2872	0.7769
children	7754.6	1.2896	0.7755
education0	7865.5	1.2714	0.7865
education1	7813.8	1.2798	0.7814
education2	7730.3	1.2936	0.7730
religion	8244.8	1.2129	0.8245
TV	7759.5	1.2887	0.7760

One can use autocorrelation and trace plots to examine the mixing of a Markov-Chain. A closely related measure of mixing is the effective sample size (ESS) (Kass et al. 1998). Low ESS indicates bad mixing of the Markov-Chain. *Effective sample size* is a measure of how well a Markov-Chain is mixing. An effective sample size that is substantially lower than the actual sample size shows slow or poor mixing of the Markov-Chain.

15.6.2 Data Analysis using R

To analyze this data using R, we first load the data into R session using the following R code chunk:

```
> # load the data
> Bangladesh = read.csv("new-Bangladesh.csv", header=T)
> # change the data type for "Education" and "CUC" to factor
> Bangladesh$education = as.factor(Bangladesh$education)
> Bangladesh$CUC = as.factor(Bangladesh$CUC)
```

The following R code replicates the standard logistic regression results from Section 8.4. with the following R code chunk. Note that R is to model the probability of CUC=1 as default, so the parameter estimates in R here and SAS above are different in sign.

```
> # call "glm" for logistic regression
>glm.Bangladesh=glm(CUC~age+education+TV+religion+children,
                   family=binomial((link='logit')), data=Bangladesh)
># print the summary of the logistic regression
>summary(glm.Bangladesh)
```

For Bayesian analysis, we call R function "MCMClogit" from "MCMCpack" to do Bayesian logistic regression using the following R code chunk:

```
>MCMCglm.Bangladesh = MCMClogit(CUC ~ age + education + TV +
        religion + children, data=Bangladesh)
># print the Bayesian logistic regression
>summary(MCMCglm.Bangladesh)
Iterations = 1001:11000
Thinning interval = 1
Number of chains = 1
Sample size per chain = 10000
1. Empirical mean and standard deviation for each variable,
   plus standard error of the mean:
                 Mean       SD       Naive SE  Time-series SE
(Intercept)     0.55788  0.090980 9.098e-04    0.0051048
age            -0.04805  0.002376 2.376e-05    0.0001297
education1      0.18264  0.044460 4.446e-04    0.0025069
education2      0.18473  0.046808 4.681e-04    0.0025751
education3      0.67528  0.072914 7.291e-04    0.0039721
TV              0.13414  0.034832 3.483e-04    0.0017733
religion        0.33040  0.047039 4.704e-04    0.0025828
children        0.31629  0.014336 1.434e-04    0.0007626
2. Quantiles for each variable:
                 2.5%      25%       50%       75%      97.5%
(Intercept)    0.38454  0.49268  0.55713  0.62085  0.73602
age           -0.05263 -0.04967 -0.04794 -0.04646 -0.04349
education1     0.10031  0.15344  0.18409  0.21210  0.27417
education2     0.09327  0.15340  0.18385  0.21698  0.27800
education3     0.53895  0.62454  0.67401  0.72393  0.81801
TV             0.05706  0.11148  0.13402  0.15689  0.19967
religion       0.24269  0.29619  0.33121  0.36194  0.42330
children       0.28761  0.30715  0.31619  0.32599  0.34483
```

The results from the Bayesian logistic regression is similar to the standard logistic regression using "glm". This is due to noninformative priors used in this exercise. If prior information was available for inclusion, options in "MCMClogit" are modified to include this prior information, which would change the results.

15.7 Summary and Comments

In this chapter, we covered the basic concepts of the Bayesian paradigm for analyzing continuous and binary data, obtaining parameter estimates, and making inferences. We demonstrated Bayesian modeling using data from Chapter 5 (Wilson et al. 2023) with multiple regression models for continuous data and data from Chapter 7 (Wilson et al. 2023) with logistic regression models for binary data. Both SAS and R were used to obtain the results. In biomedical studies and social interventions, existing historical data are often available and can be valuable for constructing Bayesian priors. When historical data are rich and relevant, more informative or less diffuse priors can be derived, enhancing the efficiency of clinical trials and other research studies. Even when historical data are limited, less informative or more diffuse priors can be used, ensuring that the Bayesian paradigm remains applicable.

15.8 Bayesian Analysis: Advantages and Disadvantages

There are both advantages and disadvantages, as well as differences and similarities, between Bayesian methods and classical frequentist methods. When the sample size is large for parametric models, Bayesian inference and frequentist methods often yield similar results. However, Bayesian analysis offers several distinct advantages:

1. **Incorporation of Prior Information:** Bayesian analysis combines prior information with current data within a robust decision-theoretical framework. This allows for the incorporation of past information about a parameter, presenting a prior distribution for future analysis. When new data become available, the previous posterior distribution can be used as a new prior.
2. **Exact Inferences:** Bayesian analysis provides exact inferences conditional on the data, without relying on asymptotic theory. This means that inferences are handled the same way for both small and large sample sizes.
3. **Adherence to the Likelihood Principle:** Bayesian analysis strictly follows the likelihood principle, unlike classical inference, which does not always adhere to this principle.
4. **Interpretable Results:** Bayesian analysis offers interpretable answers, such as stating that "the true parameter has a probability of 0.95 of falling within a 95% credible interval."

5. **Versatility:** Bayesian analysis provides a convenient framework for a wide range of models, including hierarchical models and problems involving missing data.

Despite these advantages, Bayesian analysis also has some disadvantages:

1. **Selection of Priors:** Bayesian analysis does not provide a direct method for selecting priors. There is no unified approach to choosing a prior, which often necessitates sensitivity analysis.
2. **Influence of Priors:** The resulting posterior distributions can be heavily influenced by the choice of priors, which may affect the analysis.
3. **Computational Cost:** Bayesian analysis can be computationally intensive, especially for models with a large number of parameters. Additionally, simulations may yield slightly different results unless the same random seed is used (Berger 1985; Wasserman 2004).

15.9 Exercises

Using the Bangladesh data with CUC as the response, fit a Bayesian logistic regression model with age, children, education, religion, home, and radio as covariates:

1. Is age significant in the model?
2. Is having a radio in the home a significant factor in contraceptive use?
3. Do these variables differ from those in the model fitted earlier in this chapter?
4. Do the results differ from the results based on the predictors used in this chapter?

References

Berger, J. O. (1985). *Statistical Decision Theory and Bayesian Analysis*, 2nd ed. New York: Springer-Verlag.

Brooks, S., Gelman, A., Jones, L., and Meng, X. (2011). *Handbook of Markov-Chain Monte-Carlo*. Boca Raton, FL: Chapman & Hall/CRC Handbooks of Modern Statistical Methods.

Brooks, S. P. and Roberts, G. O. (1998). Assessing convergence of Markov-Chain Monte-Carlo algorithms. *Statistics and Computing*, *8*, 319–335.

Carlin, B. P. and Louis, T. A. (2008). *Bayesian Methods for Data Analysis*, 3rd ed. Boca Raton, FL: Chapman & Hall/CRC.

Chen, D. G. and Fraser, M. W. (2017). A Bayesian perspective on intervention research: Using prior information in the development of social and health programs. *Journal of the Society for Social Work and Research*, *8*(3), 441–456.

Chen, D. G., Peace, K. E., and Zhang, P. (2017). *Clinical Trial Data Analysis Using R/SAS*. Boca Raton: Chapman & Hall/CRC Biostatistics Series.

Cowles, M. K. and Carlin, B. P. (1996). Markov-Chain Monte-Carlo convergence diagnostics: A comparative review. *Journal of American Statistical Association*, *91*(434), 883–904.

Gamerman, D. and Lopes, H. F. (2006). *Markov-Chain Monte-Carlo: Stochastic Simulation for Bayesian Inference*, 2nd ed. Boca Raton, FL: Chapman & Hall/CRC Texts in Statistical Science.

Gelman, A., Carlin, J. B., Stern, H. S., Dunson, D. B., Vehtari, A., and Rubin, D. B. (2013). *Bayesian Data Analysis*, 3rd ed. Boca Raton, FL: Chapman & Hall/CRC Texts in Statistical Science.

Kass, R. E., Carlin, B. P., Gelman, A., and Neal, R. (1998). Markov-Chain Monte-Carlo in practice: A roundtable discussion. *American Statistician*, *52*, 93–100.

Metropolis, N. (1987). The beginning of the Monte-Carlo method. Los Alamos Science (1987 Special Issue dedicated to Stanislaw Ulam), pp. 125–130.

Wasserman, L. (2004). *All of Statistics: A Concise Course in Statistical Inference*. New York: Springer-Verlag.

Wilson, J. R., Chen, D. G., and Peace, K. E. (2023). *Statistical Analytics for Health Data Science with SAS and R*. Boca Raton, FL: Chapman and Hall/CRC Biostatistics Series.

16

Jointly Modeling to Analyze Longitudinal and Survival Data with Bayesian Approach

16.1 Research Interest/Question

In public health studies, it is common to collect different types of outcome data from each patient to address complex, time-related research questions. A typical example involves assessing the trajectories of time-varying biomarkers (such as CD4 counts in HIV patients over multiple years) and their relationship to the timing of specific events of interest (e.g., death due to AIDS or dropout from the study). Traditionally, these different outcomes are often analyzed separately: longitudinal trends are estimated using mixed-effects models (such as random-intercept and random-slope models) as discussed in Chapters 9, 10, and 11 of Wilson et al. (2023), while time-to-event data (e.g., AIDS-related death or censoring due to dropout or study termination) are analyzed using Cox-type survival models, as covered in Chapter 13 (Brombin et al. 2016; Roustaei et al. 2018).

However, this traditional approach of analyzing the data independently has at least three potential limitations. First, separate analyses of the responses fail to reflect the interconnectedness of real-world scenarios, as they assume that HIV patients' CD4 trajectories evolve in isolation from eventual AIDS-related death or censoring. Second, analyzing a single outcome independently risks information loss from the other outcome, leading to biased and inefficient estimation of intervention effectiveness (Ibrahim et al. 2010). Third, longitudinal outcome analyses may inaccurately handle missing values when they do not incorporate available time-to-event data (Brombin et al. 2016). An optimal solution to these challenges is to model these responses simultaneously trend of using a joint-modeling approach. This approach produces more efficient estimates of treatment and intervention effectiveness.

This chapter aims to further stimulate interest in joint modeling and promote its application by providing an overview of the approach and its advantages. We will demonstrate how to apply joint modeling to analyze data from public health clinical trials and intervention studies where

DOI: 10.1201/9781003596370-5

both longitudinal and time-to-event data are available. To illustrate this in a real-world public health setting, we will reanalyze classical data from an HIV/AIDS study that has been widely used by other researchers, including Goldman et al. (1996) and Guo and Carlin (2004). This data are publicly available through the R packages "JM" and "JMBayes".

16.2 HIV/AIDS Data and Preliminary Data Analysis

In this study, both longitudinal data and survival data were collected to compare the efficacy and safety of two antiretroviral drugs in treating patients who had failed or were intolerant of zidovudine (AZT) therapy. In this dataset, 467 HIV-infected patients met entry criteria (either an AIDS diagnosis or two CD4 counts of 300 or fewer, and fulfilling specific criteria for AZT intolerance or failure). The patients were randomly assigned to receive either didanosine (ddI) or zalcitabine (ddC). CD4 counts were recorded at study entry and again at the 2-, 6-, 12-, and 18-month visits. The times to death were also recorded.

The longitudinal CD4 data is available in R library *JMbayes* as a built-in data frame named *"aids"* with 1,408 observations on the following nine variables. The nine variables are (1) "patient" as patients identifier for a total of 467 patients, (2) "Time" as the time-to-death or censoring, (3) "death" as censoring indicator with 0 denoting censoring and 1 death, (4) "CD4" as the CD4 measure which is square root transformed CD4 cells count to linearize the relationship between the CD4 measures and the associated "obstime", (5) "obstime" as the time points at which the CD4 cells count was recorded, (6) "drug" as a factor with levels *ddC* denoting zalcitabine and *ddI* denoting didanosine, (7) "gender" as a factor with levels *female* and *male*, (8) "prevOI" as a factor with levels *AIDS* denoting previous opportunistic infection (AIDS diagnosis) at study entry, and *noAIDS* denoting no previous infection, (9) "AZT" as a factor with levels *intolerance* and *failure* denoting AZT intolerance and AZT failure, respectively. Table 16.1 provides a sample of the data format from the first three patients with ten measures. For example, the first patient was censored (i.e., not dead with death = 0) at "Time" of 16.97 months. Patient #1 was observed at baseline, 6 months, and 12 months with the square root of CD4 counts of 10.68, 8.43, and 9.43, respectively. Patient #1 (gender is "male") was randomly assigned to "drug" of zalcitabine (i.e., ddC). Patient #1 had a previous opportunistic infection ("prevOI"), diagnosed with AIDS at study entry, and was AZT *intolerance.*

For survival analysis, the first row of each patient in this dataset was selected to build a new data frame *"aids.id"*, which has 467 rows for the 467 patients.

TABLE 16.1

Data Format from the First Three Patients

Patient	Time	Death	CD4	obstime	Drug	Gender	prevOI	AZT
1	16.97	0	10.68	0	ddC	Male	AIDS	Intolerance
1	16.97	0	8.43	6	ddC	Male	AIDS	Intolerance
1	16.97	0	9.43	12	ddC	Male	AIDS	Intolerance
2	19.00	0	6.32	0	ddI	Male	noAIDS	Intolerance
2	19.00	0	8.12	6	ddI	Male	noAIDS	Intolerance
2	19.00	0	4.58	12	ddI	Male	noAIDS	Intolerance
2	19.00	0	5.00	18	ddI	Male	noAIDS	Intolerance
3	18.53	1	3.46	0	ddI	Female	AIDS	Intolerance
3	18.53	1	3.61	2	ddI	Female	AIDS	Intolerance
3	18.53	1	6.16	6	ddI	Female	AIDS	Intolerance

We illustrate the longitudinal trend of the square-rooted CD4 measure by plotting a line graph for each of the 467 patients, as shown in Figure 16.1. In this figure, the baseline CD4 measures range from 0 to 20, with a similar range observed at the 2-, 6-, 12-, and 18-month visits (i.e., "obstime"). An examination of the trend for each patient reveals a reasonable linear relationship between CD4 measure and "obstime". These linear relationships suggest different intercepts and slopes across the 467 patients, indicating the suitability of a linear mixed effects (LMEs) model with random intercepts and random slopes for analyzing the CD4 longitudinal data, as demonstrated in Chapter 10 of Wilson et al. (2023).

Additionally, there are missing values due to death and dropouts. Data are available for 467 patients at baseline, 368 patients at 2 months, 310 patients at 6 months, 226 patients at 12 months, and only 34 patients at 18 months. These missing values are mostly missing not at random (MNAR) due to death and dropout, meaning that typical multiple imputation techniques cannot be directly applied. Therefore, these longitudinal data should be analyzed in conjunction with the survival model (i.e., "time" to death or censoring) within a joint-modeling framework.

16.3 Statistical Joint Modeling

The development of joint modeling is designed to link time-dependent longitudinal observations (i.e., CD4 counts) with time-to-event outcomes (i.e., time-to-death due to AIDS) to simultaneously assess drug effects in HIV patients. The motivation behind joint modeling is to produce more efficient estimates than those obtained from separately modeling continuous

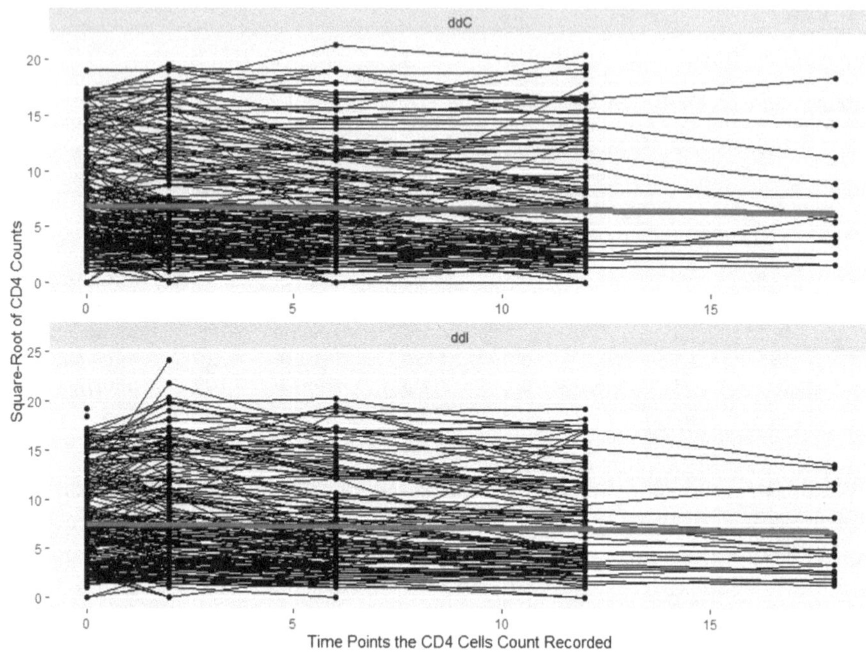

FIGURE 16.1
Longitudinal Trends for CD4 Measures (Square Root-Transformed CD4 Counts) for the Two Drug Treatments.

responses or time-to-event outcomes. This is particularly important as the longitudinal observations are time-dependent, measured with error, and subject to MNAR dropouts.

In the following subsections, we illustrate the use of joint modeling with AIDS data by presenting three competing models: (a) separate analysis of longitudinal continuous data on CD4 measures, (b) separate analysis of survival data on time-to-death due to HIV/AIDS, and (c) joint modeling of both the longitudinal continuous data and survival data. We then present and compare the results of these three models to highlight the superiority of the joint model over the classical approaches in handling outcome-dependent missingness and retaining both longitudinal continuous and survival data.

16.3.1 Separate Modeling of Longitudinal Continuous Data

The trend of the CD4 over time trajectories consists of n = 467 in the AIDS study. As discussed in Chapter 9 (Wilson et al 2023), the observed response for patient i at time t is denoted by $y_i(t)$, $i = 1, \ldots, n$; and $t = 1, \ldots, T_i$. These observed responses are continuous in nature where

$$y_i(t) = m_i(t) + \epsilon_i(t),$$

with $m_i(t)$ represents the true mean at time *t* for patient *i*, and $\epsilon_i(t)$ represents the random error, which is assumed to be $N(0, \sigma^2)$. The true mean, $m_i(t)$, is defined in a LMEs model as $x_i'(t)\beta + z_i'(t)b_i$ where

$$y_i(t) = m_i(t) + \epsilon_i(t) = x_i'(t)\beta + z_i'(t)b_i + \epsilon_i(t) \tag{16.1}$$

In Equation 16.1, $x_i'(t)\beta$ is the fixed-effects component linked to the unknown fixed-effects parameter, β (e.g., such as drug effectiveness) and $z_i'(t)b_i$ is the between-patient random-effects component with a parameter of b_i. This mixed-effects model is common among public health researchers who model longitudinal continuous data. Such mixed-effects models are available in several R packages, primarily *nlme* (Pinheiro et al. 2009) and *lme4* (Bates and Maechler 2010). Extensive literature exists on model specifications and parameter estimation in mixed-effects models using the maximum likelihood estimation (Guo and Carlin 2004; Henderson et al. 2000; Hsieh et al. 2006). The syntax used to compute the LMEs model in the R statistical programming language and computing environment is as follows:

```
R> lmeFit.aids = lme(CD4 ~ obstime + drug, random = ~ obstime
   |patient, data = aids)
R> summary(lmeFit.aids)
```

16.3.2 Separate Modeling of Time-to-Event Data

Continuing the modeling of the time-to-event data (time-to-death due to AIDS) in Chapter 13, let T_i denote the observed time ("Time" in Table 16.1) for the *i*th patient; δ_i ("death" in Table 16.1) is the censoring indicator that takes the value 1 if the patient is dead and 0 if otherwise. Therefore, the observed time-to-event data consist of the pairs $\{(T_i, \delta_i), i = 1, \ldots n\}$. The typical statistical approach to modeling time-to-event data is to use survival model with Cox proportional hazards regression (Cox 1972), where the hazard function [*h*(*t*)] consist of two parts. The first part focuses on the underlying baseline hazard function $\left[h_0(t)\right]$ to describe how the risk of event (death due to AIDS) per time unit changes over time at baseline levels of the covariates. The second part of the effect parameters, γ, describes how the hazard varies in response to explanatory covariates, w, (drug effect and other covariates, etc.) where

$$h(t \mid w) = h_0(t) exp(w'\gamma) \tag{16.2}$$

The computation around this model can be obtained in R with the *survival* package (Therneau and Lumley 2009). The R code used for this Cox proportional hazards regression is

```
R> survFit.aids = coxph(Surv(Time, death) ~ drug, x = TRUE, data = aids.id)
R> summary(survFit.aids)
```

16.3.3 Joint Modeling of Longitudinal Continuous Data and Time-to-Event Data

A joint model is proposed to address the effect on CD4 data simultaneous with the time-to-death due to HIV/AIDS. The joint modeling consist of the longitudinal continuous outcome, $y_i(t)$, for patient i at time t in Equation (16.1) and the time-to-event data, (T_i, δ_i), based on the hazard ratio in Equation (16.2). The standard approach to achieve this joint modeling is to extend the hazard risk model in Equation (16.2) to incorporate the history of longitudinal outcome from the time of enrollment to the time the child achieves permanency, which is denoted as $m(t) = \{m(u), 0 \le u < t\}$ (Therneau and Grambsch 2000). Thus, the full joint model is

$$h(t \mid m(t), w) = h_0(t) exp(w'\gamma + \alpha m_i(t)) \tag{16.3}$$

The extra parameter α in Equation (16.3) quantifies the effect of the underlying longitudinal CD4 outcome on the risk of death for every additional unit increase in the CD4 measure. In the R program output, the α parameter is labeled as *Assoct*.

Parameter estimation on this joint modeling can be obtained using a semiparametric maximum likelihood estimation method (Hsieh et al. 2006; Henderson et al. 2000; Wulfsohn and Tsiatis 1997) and implemented by Rizopoulos' (2010, 2012) with the R *JM* package.

Continuing from Chapter 15, a Bayesian perspective to estimate the parameters in the joint modeling was implemented by Rizopoulos (2016) in his R package *JMbayes* for fitting the joint models under a Bayesian approach using Markov Chain Monte Carlo (MCMC) algorithms. This R package is useful for fitting a wide range of joint models for both longitudinal continuous and categorical responses. Under the Bayesian approach, the posterior distribution of the model parameters is derived under the assumptions that given the random effects both the longitudinal and event time process are assumed independent, and the longitudinal responses of each subject are assumed to be independent. The detailed theoretical derivation is illustrated by Rizopoulos (2016). In this chapter, we use the *JMbayes* package and the following syntax to demonstrate the joint model for the CD4 longitudinal continuous responses and the time-to-death for HIV/AIDS response for patients:

```
R>  jointFit.aids <- jointModelBayes(lmeFit.aids,
    survFit.aids, timeVar = "obstime")
R>  summary(jointFit.aids)
```

This R code run MCMC for 20,000 iterations but can be changed to any other number of iterations for better convergence for the joint model with the R objects from executing the R codes for the LMEs (i.e., *lmeFit.aids*) and Cox proportional hazards models (i.e., *survFit.aids*).

16.4 Data Analysis Results

16.4.1 Results from Separate LMEs Model on CD4 Longitudinal Data

To analyze the longitudinal continuous CD4 data, an examination of the graphical relation between the CD4 measure and the number of months was made. Figure 16.1 shows that the CD4 measures have no easily determined pattern: Some patients show increasing CD4 measures, some show decreasing trends, and others exhibit no clear pattern. However, a linear trend seemed reasonable, so we used the LMEs model in (1) with the R package *nlme* to investigate the longitudinal CD4 trend along with the observed time in months from the baseline to 18 months.

In using model selection to identify the best longitudinal model for these data, we fit four models:

- Model 1: random intercept LME with interaction between "obstime" and "drug",,
- Model 2 random intercept LME without interaction between "obstime" and "drug",

- Model 3 random intercept and random slope LME with interaction between "obstime" and "drug", and
- Model 4 random intercept and random slope LME without interaction between "obstime" and "drug",.

As summarized in Table 16.2, we first compare the random intercept LME models (i.e., Model 1 vs. Model 2). Using a likelihood ratio chi-square test to compare these models, we found that Model 2 is not statistically significantly different from Model 1. The difference in the likelihood ratio statistic is 0.182, with a p-value of .670. However, both the AIC and BIC values for Model 2 are smaller than those for Model 1, suggesting that Model 2 is preferred based on the principle of model parsimony, as it includes fewer parameters (i.e., no interaction) among the random intercept LME models.

Similarly, among the two random slope LME models (i.e., Model 3 vs. Model 4), Model 4 is preferred due to nonsignificance of the likelihood ratio test, less AIC and BIC, as well as model parsimony as seen in Table 16.2.

To determine which model is preferred among the Models 2 and 4, we can get that the likelihood ratio statistic is 38.9373 (i.e., 2*(−3579.43 − (−3559.96) = 38.9373) with 2 degrees of freedom (i.e., 7 − 5 = 2), which gives the p-value of <.0001. This suggest that the Model 4 (i.e., random-slope LME without interaction) is a better fit than the Model 2 (i.e., random-intercept LME without interaction). In addition, the AIC and BIC for Model 4 (i.e., AIC = 7133.91 and BIC = 7170.65) are smaller than those for Model 2 (i.e., AIC = 7168.85 and BIC = 7195.09), which further confirmed that the random slope LME model without "obstime" and "drug" interaction is the best model among the four LME models to fit the longitudinal CD4 measure data.

The results for the best LME model (i.e., Model 4 in Table 16.2) for the longitudinal CD4 measure is presented in Table 16.3 (left-side, titled "*Separate linear mixed-effects model*"). With this Model 4, the estimated intercept parameter is $\hat{\beta}_0 = 6.918$ ($SE = 0.308$; $p < .001$), slope parameter for time trend is $\hat{\beta}_1 = -0.149$ ($SE = 0.015$; $p < .001$), and the slope parameter for the "drug" effect is $\hat{\beta}_2 = 0.549$ ($SE = 0.433$; $p < .205$). In summary with the separate longitudinal

TABLE 16.2

Model Selection Statistics for Longitudinal Linear Mixed-Effects Models

Model	Random Effects	Interaction	df	AIC	BIC	logLik	L.Ratio	p-value
1	Only intercept	Yes	6	7170.67	7202.16	-3579.34		
2	Only intercept	No	5	7168.85	7195.09	-3579.43	0.182	.670
3	Intercept and slope	Yes	8	7135.44	7177.42	-3559.72		
4	Intercept and slope	No	7	7133.91	7170.65	-3559.96	0.476	.490

LME modeling, we found a statistically significant overall declining trend for CD4 measure for these 467 patients. However, the "drug" effect is not statistically significant to the CD4 measures.

The mixed-effects model treats missing values from the longitudinal CD4 as missing at random. This is an unreasonable assumption as these missing values originate from the HIV/AIDS patients which could be a result of death or dropout. Those missing CD4 responses are not missing at random as we know their missing mechanism on when they died, and therefore, treating them as missing at random in this mixed-effects longitudinal model is erroneous and can produces biased estimates.

16.4.2 Results of Separate Cox Proportional Hazards Regression

Of the 467 patients, there are 237 randomly assigned to 'drug = ddC' and 230 to 'drug = ddI'. There are 88 deaths of the 237 patients in ddC and 100 deaths of the 230 patients in ddI. The time-to-death data are analyzed using Cox proportional hazard regression in Equation (16.2) to determine the drug effects. We found statistically nonsignificant with an estimated log hazard rate of 0.210 (SE = 1.234, p =0.5 1), which is presented in Table 16.3 (left side, titled as "*Separate Cox Proportional Hazards Regression model*"). The estimated survival function can be seen in Figure 16.2, and it can be seen that the estimated

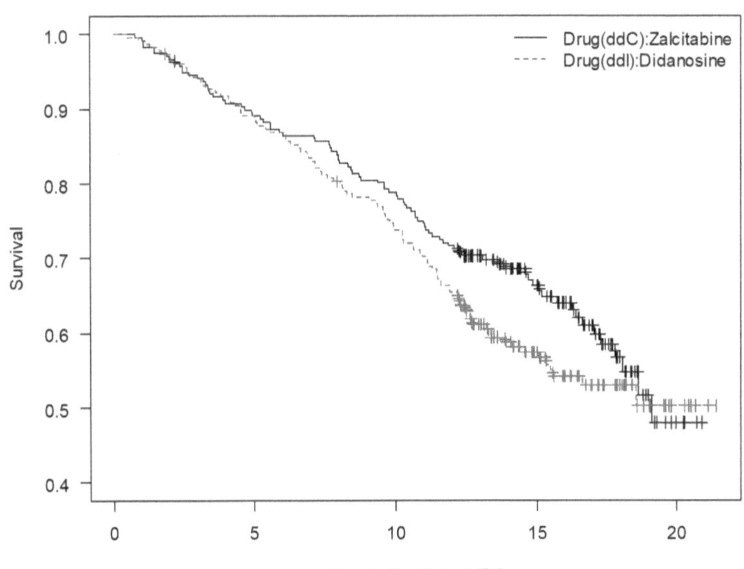

FIGURE 16.2
Estimated Survival Functions for the Two Drug Treatments.

survival functions for these two duges are very close together without much difference.

16.4.3 Results of Joint Modeling of Longitudinal CD4 and Time-to-Death

The joint modelling of longitudinal responses CD4 measure and time-to-death data involves integrating the models in (3) and (4). The model is used to test the hypothesis that the CD4 measures are associated with time-to-death, as it pertains to the type of drug.

The parameters of the joint model can be estimated using MCMC for 20,000 iterations with the R objects from the LMEs (*lmeFit.aids*) and Cox proportional hazards models (*survFit.aids*). A graphical representation of the MCMC runs for the longitudinal component of joint modeling is displayed in Figure 16.3. Similar plots are available for other parameters. Figure 16.3 demonstrates convergence in the MCMC series.

The estimated parameters for the joint longitudinal component and survival component are summarized in Table 16.3 (right side). The standard errors in the joint model are smaller than those in the separate models, suggesting that the joint model provides more efficient estimates. Consequently, the drug effect parameter estimates in the joint model are more statistically significant.

In particular, the posterior estimates in the joint model for the association parameter "Assoct" is −0.294 (SE = 0.003, $p < 0.001$) and 95% Bayesian credible interval is (−0.378, −0.207). This result provides strong evidence of an

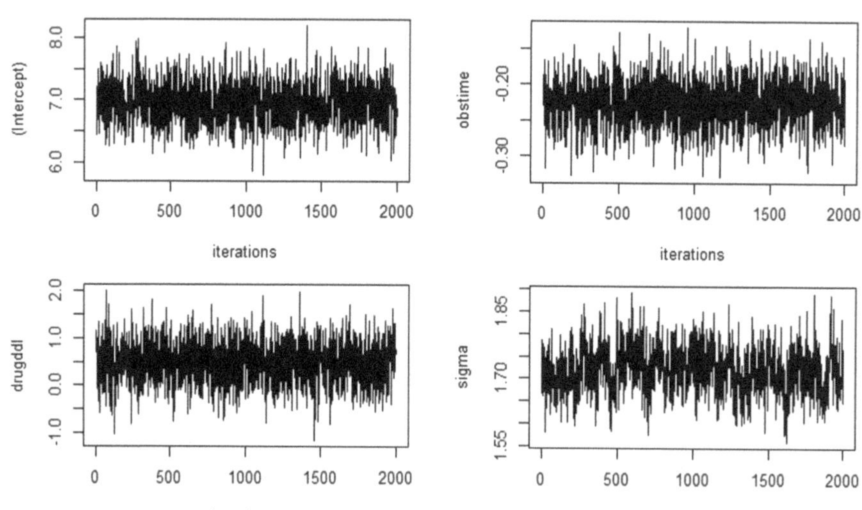

FIGURE 16.3
MCMC Illustration for the Parameters in the Longitudinal CD4 Component.

TABLE 16.3

Summary of the Results from the Two Separate Models and the Joint Models

	Separate Modeling of Longitudinal and Survival Data			**Joint Modeling**		
	Separate cox proportional hazards regression model			**Survival component**		
	Coef.	*SE*	***p*-value**	**Coef.**	*SE*	***p*-value**
Drug(ddI)	0.210	1.234	.151	0.328	0.003	.089
	Separate linear mixed-effects model			**Longitudinal component**		
	Coef.	**SE**	***p*-value**	**Coef.**	**SE**	***p*-value**
Intercept	6.918	0.308	<.001	6.949	0.008	<.001
obstime	−0.149	0.015	<0.001	−0.227	0.001	<.001
Drug(ddI)	0.549	0.433	.205	0.497	0.011	.264

association between the two submodels (longitudinal component on CD4 measure and time-to-death due to HIV/AIDS). This indicates that a patient's survival status is significantly related to the longitudinal characteristics of the CD4 which are specified the initial CD4 level and the rate of CD4 changes. This finding is clinically logical since higher CD4 counts are typically linked with better health status and patients with a more rapid decline in CD4 counts would be expected to have a poorer health condition and less survival probability.

16.5 Conclusions and Recommendations

In this chapter, data from an HIV/AIDS study are used to demonstrate the merits of joint modeling, which simultaneously handles longitudinal continuous data and time-to-event data. The joint modeling approach offers several advantages over separate modeling, both fundamentally and practically.

Fundamentally, the joint modeling framework incorporates time-to-event data into the observed longitudinal CD4 measurements, mitigating the impact of missing CD4 values due to patient death or dropout. Traditionally, in mixed-effects modeling, death or dropout in longitudinal data is often treated as missing at random, justifying the use of multiple imputations. However, this approach can be flawed in such datasets, and the failure to detect and address these errors can lead to biased estimates and, in some cases, incorrect conclusions for public health policymakers and practitioners. The joint modeling framework we present is more logically consistent with study designs that track both time-to-event and longitudinal outcome data while effectively addressing the issue of missing data.

Practically, the joint modeling framework offers higher statistical power to detect an effect or relationship when one truly exists in survival models (Brombin et al. 2016; Rizopoulos 2012; Roustaei et al. 2018). In our analysis of the HIV/AIDS data, the efficiency of the joint modeling framework was evident through the incorporation of longitudinal measures alongside the survival model. Joint modeling enhances statistical power, particularly in detecting the predictive role of CD4 trajectories in understanding the survival dynamics related to time-to-death due to AIDS. This increased statistical power has practical cost implications, as greater efficiency allows for smaller sample sizes.

Given the benefits of the joint modeling framework, we recommend that researchers designing clinical trials and public health intervention studies to understand treatment effects on time-to-event data should incorporate longitudinal data. Joint models not only handle outcome-dependent missingness effectively but also potentially allow researchers to achieve higher statistical power even with smaller sample sizes. Researchers should prioritize the collection of time-to-event data to address and statistically model missing data from longitudinal measurements. Joint modeling aligns with emerging practices in data pooling and harmonization, where Bayesian approaches are commonly used to integrate multiple data sources (Chen et al. 2017, 2020).

16.6 Exercises

Using the HIV/AIDS data in Section 16.2, repeat the analysis in Section 16.4, incorporating additional time-invariant covariates such as "gender", "PrevOI" (i.e., previous opportunistic infection and AIDS diagnosis at study entry), and "AZT" (i.e., AZT intolerance) in the longitudinal component, to answer the following research questions:

1. Is "gender" statistically significant?
2. Is "PrevOI" statistically significant?
3. Is "AZT" statistically significant?
4. Is there any statistically significant survival improvement on "drug" effect by incorporating these time-invariant covariates?

References

Bates, D. and Maechler, M. (2010). *lme4: Linear Mixed-Effects Models Using S4 Classes.* R package (Version 0.999375-34). http://CRAN.R-project.org/package=lme4

Brombin, C., Di Serio, C., and Rancoita, P. M. (2016). Joint modeling of HIV data in multicenter observational studies: A comparison among different approaches. *Statistical Methods in Medical Research, 25*(6), 2472–2487. https://doi.org/10.1177/0962280213519722.

Chen, D., Testa, M. F., Ansong, D., and Brevard, K. C. (2020). Evidence building and information accumulation: Bayesian paradigm cohesive for child welfare intervention research. *Journal of the Society for Social Work and Research, 11*(3), 1–16. https://doi.org/10.1086/711376

Chen, D. G., Ting, N., and Ho, S. (2017). Informative priors or non-informative priors? A Bayesian re-analysis of binary data from Macugen Phase III clinical trials. *Communications in Statistics-Simulation and Computation, 46*(6), 4535–4546. DOI:10.1080/03610918.2015.1122049

Cox, D. R. (1972). Regression models and life-tables. *Journal of the Royal Statistical Society: Series B (Methodological), 34*(2),187–220. https://doi.org/10.1111/j.2517-6161.1972.tb00899.x

Goldman, A., Carlin, B., Crane, L., Launer, C., Korvick, J., Deyton, L., and Abrams, D. (1996) Response of CD4+ and clinical consequences to treatment using ddI or ddC in patients with advanced HIV infection. *Journal of Acquired Immune Deficiency Syndromes and Human Retrovirology, 11*, 161–169.

Guo, X. and Carlin, B. P. (2004). Separate and joint modeling of longitudinal and event time data using standard computer packages. *The American Statistician, 58*(1), 16–24. https://doi.org/10.1198/0003130042854

Henderson, R., Diggle, P., and Dobson, A. (2000). Joint modelling of longitudinal measurements and event time data. *Biostatistics, 1*(4), 465–480. https://doi.org/10.1093/biostatistics/1.4.465

Hsieh, F., Tseng, Y. K., and Wang, J. L. (2006). Joint modeling of survival and longitudinal data: Likelihood approach revisited. *Biometrics, 62*(4), 1037–1043. https://doi.org/10.1111/j.1541-0420.2006.00570.x

Ibrahim, J. G., Chu, H., and Chen, L. M. (2010). Basic concepts and methods for joint models of longitudinal and survival data. *Journal of Clinical Oncology, 28*(16), 2796–2801. https://doi.org/10.1200/JCO.2009.25.0654

Pinheiro, J., Bates, D., DebRoy, S., and Sarkar, D. R. (2009). *nlme: Linear and nonlinear mixed effects models.* R package (Version 3.1–149). http://CRAN.R-project.org/package=nlme

Rizopoulos, D. (2010). JM: An R package for the joint modeling of longitudinal and time-to-event data. *Journal of Statistical Software, 35*(9), 1–33. http://dx.doi.org/10.18637/jss.v035.i09

Rizopoulos, D. (2012). *Joint Models for Longitudinal and Time-to-Event Data: With Applications in R.* Boca Raton, FL: Chapman & Hall/CRC Biostatistics Series.

Rizopoulos, D. (2016). The R Package JMbayes for fitting joint models for longitudinal and time-to-event data using MCMC. *Journal of Statistical Software, 72*(7): 1–46. doi:10.18637/jss.v072.i07

Roustaei, N., Ayatollahi, S. M. T., and Zare, N. (2018). A proposed approach for joint modeling of the longitudinal and time-to-event data in heterogeneous populations: An application to HIV/AIDS's disease. *BioMed Research International, 2018,* Article ID 3501238. https://doi.org/10.1155/2018/3501238

Therneau, T. and Grambsch, P. (2000). *Modeling Survival Data: Extending the Cox Model.* New York: Springer-Verlag.

Therneau, T. and Lumley, T. (2009). *"Survival": Survival analysis including penalised likelihood.* R package (Version 2.35–8). http://CRAN.R-project.org/package=survival

Wilson, J. R., Chen, D. G., & Peace, K. E. (2023). *Statistical Analytics for Health Data Science with SAS and R.* Boca Raton, FL: Chapman and Hall/CRC.

Wulfsohn, M. S. and Tsiatis, A. A. (1997). A joint model for survival and longitudinal data measured with error. *Biometrics, 53*(1), 330–339. https://doi.org/10.2307/2533118

17

Nonlinear Regression

17.1 Research Interest/Questions

It is common for researchers to believe that their research questions can be addressed with linear regression models. This assumption is valid if the problem is truly linear or if you are restricted to studying only a small subset of the population of data. Other researchers often transform their problems appropriately to fit linear models. The linear regression model assumes a linear relationship between the response and predictors. However, in some cases, the true relationship between the response and the predictors may be nonlinear. It may not make physical sense after the regression coefficients have been through a transformation. However, complexity or convenience should not be the primary focus. Therefore, while linear models are useful for much research, the fact is that nonlinearity is present in our existence and should not be avoided. This chapter concentrates on the estimation of nonlinear models. Estimation of nonlinear models usually requires finding the minimum (or maximum) of a nonlinear function. Linear model is a special case of nonlinear model. However, nonlinear regression models [A model is nonlinear if it is not linear in at least one parameter.] are important tools because there are many settings better represented by nonlinear than linear models.

17.2 Chlorine Data

As a nonlinear regression example to illustrate the fitting of nonlinear models, let's consider an example discussed in Draper and Smith's regression book (Draper and Smith, 1981, 1998). The nonlinear model uses to explain Y in terms of X has two parameters α and β and error term ε is denoted by

DOI: 10.1201/9781003596370-6

TABLE 17.1

Subset of Data

Obs	Row	X	Y
1	1	8	0.49
2	23	22	0.41
3	2	8	0.49
4	24	22	0.40
5	3	10	0.48
6	25	24	0.42
7	4	10	0.47
8	26	24	0.40

$$Y = \alpha + (0.49 - \alpha)e^{(-\beta(X-8))} + \varepsilon$$

Data first reported in Draper and Smith (1981) are considered. The response variable (*Y*) is the proportion of available chlorine in a certain quantity of chlorine solution, and the predictor (*X*) is the length of time in weeks since the product was produced. A subset of the 44 data points in the dataset is listed (Table 17.1).

17.3 Nonlinear Regression Models

The complexity associated with nonlinear regression models can be compensated by their advantages, including fewer parameters, improved interpretability, and better predictive performance (Bates and Watts 2008). A *linear* model is usually a good first approximation, but occasionally, one may need to use more complex models. Nonlinear regression models are complex as they are not linear in the parameters. Examples of nonlinear equations are as follows:

Consider three common examples of nonlinear models:

Exponential growth model $Y_i = \theta_1 e^{\theta_2 t_i} + \varepsilon_i$

Gompertz growth model $Y_i = \theta_1 e^{\{-\theta_2 e^{\theta_3 t_i}\}} + \varepsilon_i$

Weibull model $Y_i = \theta_1 e^{-(X_i/\theta_2)^{\theta_3}} + \varepsilon_i$

where Y_i denotes the response variable, t_i and X_i are explanatory variables, θ is a rate constant (parameter), and ε_i is the error term.

A key advantage of using nonlinear regression models is that their predictions are more reliable than high-degree polynomial models, especially outside the range of data (i.e., extrapolation). While the main disadvantage of nonlinear models is that they can be less flexible than linear models, and that generally, there is no analytical solution for estimating the parameters.

More importantly, the choice of the model is crucial. The choice of the model is determined in different ways including trial and error or theory or past studies. This discussion is further complicated by the fact that one needs a model that will converge and supply parameter estimates. Moreover, convergence of models depends on the starting values in the iterative numerical method.

17.4 Fitting Nonlinear Models

It is not uncommon when fitting a linear model, the general form of the model may be unknown. It is necessary to identify the dependent and independent variables and the terms to include in the model. For a nonlinear model, the form of the model must be specified, the parameters to be estimated identified, and starting values for those parameters must be provided. In a linear model, parameter estimates are obtained using specific equations that provide a unique solution minimizing the sum of squared errors. However, in the case of a nonlinear model, iterative techniques are generally used to converge on to a solution that provides the best model fit. It is important for the analyst to ensure that convergence has been attained. There are certain conditions, which can prevent convergence.

These conditions include as follows:

- Incorrect specification of the model

  This is the case when the data and the model do not exhibit the same relationship:
- Poor initial starting values for the parameters

  Poor initial starting values can cause lack of convergence or convergence to a local minimum
- Overdefined model

  This is the case when a model has more parameters than are necessary for the relationship
- Insufficient iterations

  Convergence is not obtained before the maximum number of iterations is reached.
- Insufficient data

  Not enough data to generate a model.

The fit of a nonlinear model is not a single-step procedure but an involved process that requires careful examination of each individual step. Consider the use of statistical programs, SAS, R, SPSS, and STATA to fit these nonlinear models. Depending on the objective and the application of the problem, different priorities are set when fitting nonlinear models. These include obtaining acceptable parameter estimates and a good model fit while meeting standard assumptions of statistical models. Researchers have systematically presented seven steps:

(i) Choose candidate models,
(ii) Set starting values; use theory; use graphical display; take a transformation and get a rough estimate; and use grid search
(iii) Fit models,
(iv) Check convergence and parameter estimates,
(v) Find the "best" model among competing models,
(vi) Check model assumptions (residual analysis), and
(vii) Calculate statistical descriptors and confidence intervals. The associated feedback mechanisms are also addressed (i.e., model variance homogeneity).

The most critical step that distinguishes nonlinear models from linear models is that the choice of the main function is critical and this can be difficult without appropriate guidance. Usually, nonlinear regression model is used to estimate the parameters in a nonlinear model without performing hypothesis tests. In this case, the usual assumption about the normality of the errors is not needed. Instead, the main assumption is that the data are represented by the model.

17.4.1 Normal Probability Plot

If the errors are normally distributed, the data points on the normal probability plot will fall along a straight line. Major deviations from this straight line reflect departures from normality. Large deviations from the normal probability plot straight line indicate outliers. Curvature at both ends of the plot indicates long or short distributional tails, convex or concave curvature indicates a lack of symmetry, and gaps, plateaus, or segmentation in the normal probability plot may require a closer examination of the data or model. It is not recommended to use this diagnostic with small sample sizes.

17.4.2 Residual versus Predicted Plot

This is a scatterplot of the residuals versus each covariate. The preferred pattern is a rectangular shape or point cloud. Any nonrandom pattern may require a redefining of the model. A plot is always helpful. One looks for patterns. Certain pattern is an indicator of nonconstant variance, while sloping or curved band signifies inadequate specification of the model. A sloping band with increasing or decreasing variability suggests nonconstant variance and inadequate specification of the model.

17.4.3 Prediction

One uses the model to predict Y for new values of the covariates. Most programs allow one to include the covariate values and put "." for the response Y.

17.5 Data Analysis with SAS and R

In this example, the model was suggested based by a subject matter expert containing two unknowns:

1. The asymptotic baseline value reached at large values of weeks, and
2. The exponential rate of decay. The steps of reporting follow Al-Kaisi et al. (2015).

Data:
In the chlorine data, the response variable (Y) is the proportion of available chlorine in a certain quantity of chlorine solution, and the predictor (X) is the length of time in weeks since the product was produced (Draper and Smith 1981). Initially, the proportion of chlorine is 0.50. Over 8 weeks, the proportion declines to 0.49.

Choose candidate model:
Based on the background information in the hypothesized model for predicting Y from X

$$y = \alpha + (0.49 - \alpha)e^{(-\beta(x-8))}$$

The parameters have clear meaning and are very suitable to answer the research questions. The problem is to estimate the parameters α and β of the nonlinear model.

Starting values:
The biggest challenge with the estimation in nonlinear model is the need to supply starting values for the parameters. The convergence of the algorithms depends on a great deal upon supplying appropriate starting values. One tries to fit the nearest linear model to help provide starting values. In this case, consider $y = \alpha - \beta(x-8)$ with two unknown parameters $[\boldsymbol{\alpha}, \boldsymbol{\beta}]$. The SAS and R are used to fit this model.

17.5.1 DATA Analysis with SAS Program

```
Data Chap17;
input Row X Y@@;
datalines;
1   8 0.  49  23  22  0.41
2   8 0.  49  24  22  0.40
3   10 0. 48  25  24  0.42
4   10 0. 47  26  24  0.40
5   10 0. 48  27  24  0.40
6   10 0. 47  28  26  0.41
.................................................
22  22 0. 41  44  42  0.39
;
* Y= α +(0.49- α)exp(-β (X-8)+ ε;
proc reg;
model y=x; run;
```

Fit a linear regression model to get an idea of possible starting values. Starting values are essential.

```
proc nlmixed data=chap17 corr;
parms α =1 β =-0.01  s2e=1;
model Chlorine= α +((0.49- α)*exp(-β *(X-8)));
Model Chlorine ~ NORMAL(y, s2e);
Predict Model Y out=Pred;
proc print; run;
```

Underlined words are SAS. One can use this program to mimic ones model. The fit of nonlinear model is done with PROC NLMIXED.

Parameter Estimates

Variable	DF	Parameter Estimate	Standard Error	t Value	Pr > |t|
Intercept	1	0.48551	0.00589	82.42	<.0001
X	1	-0.00272	0.00024311	-11.17	<.0001

The linear regression model suggests good starting values are α = 0.4855 and β =-0.00272.

The NLMIXED Procedure

Specifications	
Data Set	WORK.CHAP17
Dependent Variable	Chlorine
Distribution for Dependent Variable	Normal
Optimization Technique	Dual Quasi-Newton
Integration Method	None

The technique used to obtain the parameter estimates for the nonlinear model is based on iterative method.

Dimensions	
Observations Used	44
Observations Not Used	0
Total Observations	44
Parameters	3

Parameters

α	β	s2e	NegLogLike
1	-0.01	1	40.4665643

There are three parameters, α, β, and assumed variance of the distribution of Y. The starting values are α =1 β =-0.01 based on the simple linear regression model and variance =1.

Iteration History

Iter	Calls	NegLogLike	Diff	MaxGrad	Slope
1	6	35.4606278	5.005936	57.68942	-166.145
2	11	31.3310177	4.12961	147.4001	-1837.13
3	15	28.1414143	3.189603	243.7816	-252.944
29	78	-137.37447	0.000017	6.04182	-0.00003
30	80	-137.37447	2.448E-8	0.147081	-4.03E-8

NOTE: GCONV convergence criterion satisfied.

The iterative procedure converged based on certain stopping criterion.

Fit Statistics	
-2 Log Likelihood	-274.7
AIC (smaller is better)	-268.7
AICC (smaller is better)	-268.1
BIC (smaller is better)	-263.4

Parameter Estimates

Parameter	Estimate	Standard Error	DF	t-Value	Pr>|t|	Alpha	Lower	Upper	Gradient
α	0.3901	0.004974	44	78.44	<.0001	0.05	0.3801	0.4002	0.00263
β	0.1016	0.01320	44	7.70	<.0001	0.05	0.07502	0.1282	-0.00085
s2e	0.000114	0.000024	44	4.67	<.0001	0.05	0.000065	0.000163	-0.14708

At convergence $\hat{\alpha}$=0.3901 with standard error 0.00497. $\hat{\beta}$=0.1016 with standard error 0.01320. Both α and β are significant in the model with p<.0001. The variance is also significant thus, *t*

$$\hat{Y} = 0.390 \; +(0.49\text{-} \; 0.390\,)\, e^{-0.102\,(X-8)}$$

Correlation Matrix of Parameter Estimates

Row	Parameter	α	β	s2e
1	A	1.0000	0.8900	0.000044
2	B	0.8900	1.0000	0.000040
3	s2e	0.000044	0.000040	1.0000

This matrix displays the asymptotic correlations of the parameter estimates. When these correlations are high (absolute value greater than 0.95), the precision of the parameter estimates is suspect

Obs	Row	X	Y	Pred	StdErrPred	DF	tValue	Probt	Alpha	Lower	Upper
6	25	24	0.42	0.409	.001916017	44	213.87	4.98E-68	0.05	0.405	0.413
7	4	10	0.47	0.471	.001401222	44	336.58	1.09E-76	0.05	0.468	0.474
8	26	24	0.40	0.409	.001916017	44	213.87	4.98E-68	0.05	0.405	0.413

Obs	Row	X	Y	Pred	StdErrPred	DF	tValue	Probt	Alpha	Lower	Upper
43	22	22	0.41	0.414	.002037005	44	203.34	4.58E-67	0.05	0.410	0.418
44	44	42	0.39	0.393	.003615105	44	108.79	3.88E-55	0.05	0.386	0.400

The predicted values and the standard errors are given along with their confidence intervals.

17.5.2 Data Analysis with R Program

The nonlinear regression model is fitted using R program. α, β

```
NLM<-nls(Y~ α +((0.49- α)*exp(-β *(X-8))),start=list(α =0.4, β =0.1))
summary(NLM)
##
## Formula: Y ~ α + ((0.49 - α) * exp(-β * (X - 8)))
## Parameters:
##    Estimate Std. Error t value Pr(>|t|)
## α 0.390140    0.005045  77.333  < 2e-16 ***
## β 0.101633    0.013360   7.607  1.99e-09 ***
```

At convergence $\hat{\alpha}$=0.3901 with standard error 0.00497. $\hat{\beta}$=0.1016 with standard error 0.01320. Both α and β are significant in the model with p<.0001. Their standard error are so small that they are significant estimates in the model.

$$\hat{Y} = 0.390 + (0.49 - 0.390)\, e^{-0.102(X-8)}$$

```
## ---
## Residual standard error: 0.01091 on 42 degrees of freedom
## Number of iterations to convergence: 3
## Achieved convergence tolerance: 8.464e-07
Residuals<-resid(NLM)
plot(X,Residuals,main="Residuals for Y")
abline(h=0)
```

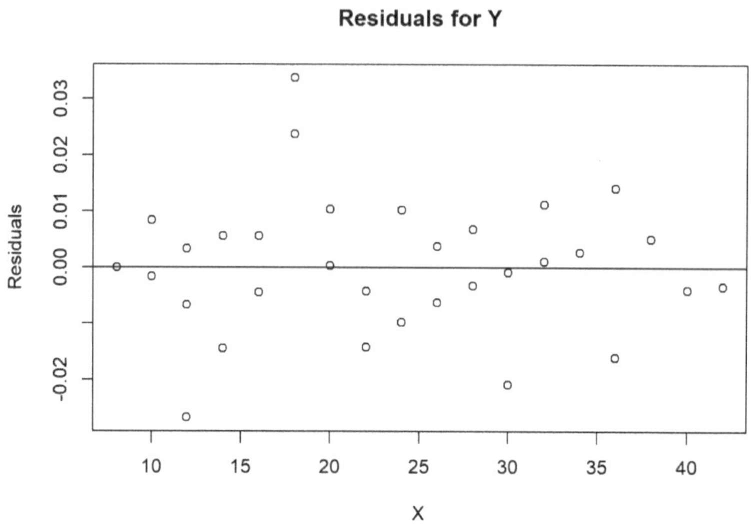

FIGURE 17.1
Residual Plot.

The above code would produce the resuidual plot in Figure 17.1:

17.6 Results and Discussions

The iterative process converges to the fitted equation, which takes the form:

$$\hat{Y} = 0.39 + (0.49 - 0.39)e^{(-0.102(x-8))}$$

At convergence A=0.3901 with standard error 0.0050. B=0.1016 with standard error 0.01336. Both A and B are significant in the model with p<.0001. The variance is also significant thus; the variation in Y is significant. . Thus the predicted nonlinear model is.

$$\hat{Y} = 0.390 \; +(0.49\text{- } 0.390\,)\; e^{-0.102\,(X-8)}$$

Comments

The model fitted in this chapter has one independent variable. If more than one independent variable is included in the model, surface and contour plots can be created. For example, Draper and Smith (1998) report on an experiment in which the fraction of material Y remaining after a chemical reaction was described by the model

17.7 Exercises

1. Use a different model say $Y = \alpha + \beta X^{-1}$ transformation of the model and compare with the results obtained.
    a. Obtain the predicted values
    b. Obtain the residuals
2. Using the model fitted in this chapter,
    a. Obtain the predicted values
    b. Obtain the residuals
    c. Compare the predicted values and residuals.
3. Compare the results with the two approaches.

References

Al-Kaisi, M. M., Archontoulis, S. V., Kwaw-Mensah, D. and Miguez, F. (2015). Tillage and crop rotation effects on corn agronomic response and economic return at seven Iowa locations. *Agronomy Journal, 107*, 1411–1424. doi:10.2134/agronj14.0470

Bates, D. and Watts, D. (2008). *Nonlinear Regression Analysis and Its Applications* (pp. 32–66). Hoboken, NJ: Wiley.

Draper, N. R. and Smith, H. (1981) *Applied Regression Analysis*. Hoboken, NJ: Wiley.

Draper, N. R. and Smith, H (1998). *Applied Regression Analysis*. 3rd edn. New York: Wiley.

18

Statistical Meta-analysis

In this chapter, we present an introduction to meta-analysis (MA) with an overview of classical fixed-effects and random-effects MA (RE-MA) models to synthesize summary statistics. We also present meta-regression to explain the between-study heterogeneity. Metadata from 13 clinical trials to assess Bacillus Calmette–Guerin vaccine in the prevention of tuberculosis are used to demonstrate the application of these meta-analysis (MA) models in R and SAS.

18.1 Research Interest/Questions

In scientific research, it is common to obtain information from different sources and then synthesize the information to draw reliable inferences and thereby arrive at robust conclusions. This process of searching for and integrating diverse information is known as a "systematic review". When it involves generating overall results based on pooled data, it is usually referred to as "meta-analysis" (MA). MA is a widely established scientific field with extensive publications and books on MA, such as Whitehead (2003), Hartung et al. (2008), Borenstein et al. (2009), Pigott (2012), Chen and Peace (2013), and Schwarzer et al. (2015), to list a few. In this chapter, we present an introduction to MA with an overview of classical fixed-effects and random-effects meta-analysis (RE-MA) models to synthesize summary statistics. We also present meta-regression to explain the between-study heterogeneity. Metadata from 13 clinical trials to assess Bacillus Calmette-Guerin (BCG) vaccine in the prevention of tuberculosis (TB) are used to demonstrate the application of these MA models in R and SAS.

DOI: 10.1201/9781003596370-7

18.2 Metadata on Bacillus Calmette-Guerin Vaccine Clinical Trials

A systematic review of 13 clinical trials evaluating the impact of a BCG vaccine in the prevention of TB is widely used to illustrate MA and meta-regression. Results are presented in Hartung et al. (2008), Borenstein et al. (2009), Chen and Peace (2013), and the R library *metafor*. The source dataset reports the results in the study of each trial investigating the effectiveness of BCG in the treatment of TB and is summarized in Table 18.1.

In Table 18.1, "Trial" is the clinical trial number from 1 to 13; "Author(year)" denotes the authorship from the 13 studies and the year published for the study; "tpos" is the number of TB positive cases in the BCG-vaccinated group; "tneg" is the number of TB negative cases in the BCG-vaccinated group; "cpos" is the number of TB positive cases in the control group; "cneg" is the number of TB negative cases in the control group; and "ablat" denotes the absolute latitude of the study location (in degrees). Originally, a RE-MA model was fitted to quantify the efficacy of the BCG vaccine against TB. The model revealed that the BCG vaccine significantly reduces the risk of TB in the presence of significant heterogeneity.

TABLE 18.1

Data from 13 Studies on BCG Vaccine for Preventing Tuberculosis

Trial	Author (year)	TPOS	TNEG	CPOS	CNEG	ABLAT
1	Aronson (1948)	4	119	11	128	44
2	Ferguson and Simes (19490	6	300	29	274	55
3	Rosenthal et al.(1960)	3	228	11	209	42
4	Hart and Sutherland (1977)	62	13536	248	12619	52
5	Frimodt-Moller et al. (1973)	33	5036	47	5761	13
6	Stein and Aronson (1953)	180	1361	372	1079	44
7	Vandiviere et al.(1973)	8	2537	10	619	19
8	TPT Madras (1980)	505	87886	499	87892	13
9	Coetzee and Berjak (1968)	29	7470	45	7232	27
10	Rosenthal et al. (1961)	17	1699	65	1600	42
11	Comstock et al. (1974)	186	50448	141	27197	18
12	Comstock and Webster (1969)	5	2493	3	2338	33
13	Comstock et al. (1976)	27	16886	29	17825	33

18.3 Meta-analysis Models: Fixed Effects and Random Effects

In this section, we provide an overview of MA. We present two models: fixed-effects MA and RE-MA. We also quantify heterogeneity using Q-statistic, τ^2 index, H index, and I^2 index. In addition, a meta-regression is used to explore the heterogeneity with study-level moderators.

18.3.1 Fixed-Effects Meta-analysis Models

A typical MA combines K independent studies each with the population effect size, δ_k $(k = 1, 2, ..., K)$. The population effect size δ_k is estimated using the observed effect size, $\hat{\delta}_k$. In the fixed-effects MA, the assumption is that there is no between-study variation among the K studies regardless of where, when, with whom, and how the studies were conducted. This is referred to as "homogeneity assumption". It assumes that the underlying population effect sizes δ_k are the same across all studies (i.e., $\delta_1 = \ldots = \delta_K = \delta$). Thus, we assume that the sample of observed study effect sizes $\hat{\delta}_k$ comes from a population with a known sampling error. Therefore, a typical fixed-effects MA model is described as

$$\hat{\delta}_k = \delta + \varepsilon_k \tag{18.1}$$

where ε_k $(k = 1, 2,\ldots, K)$ represents the within-study variation and assumed normally distributed with mean 0 and known variance σ_k^2, that is, $\varepsilon_k \sim N\left(0, \sigma_k^2\right)$. In MA, the effect size σ_k^2 for all included studies is assumed known. This differs from the assumptions in the analysis of variance or regression models, in which the error variance is unknown and is estimated from the data. These variances from K studies are calculated as dichotomized outcome variables. However, for continuous target variables or other types of variables, the variance is usually estimated from previously reported literature. The meta-estimate for the global population effect size $\hat{\delta}$ (combining data from all the independent studies) is estimated using a weighted-mean method where the weights are the inverse of the known variances. Thus,

$$\hat{\delta} = \frac{\sum_{k=1}^{K} w_k \hat{\delta}_k}{\sum_{k=1}^{K} w_k} \tag{18.2}$$

where $w_k = 1/\sigma_k^2$ $(k = 1,\ldots,K)$. The variance of $\hat{\delta}$ the global estimate of effect size,

$$Var\left(\hat{\delta}\right) = 1/\sum_{k=1}^{K} w_k \tag{18.3}$$

Thus, a 95% confidence interval for the global effect size of δ is

$$\hat{\delta} \pm 1.96\sqrt{Var\left(\hat{\delta}\right)} \tag{18.4}$$

and a test statistic for $H_0\!: \delta = 0$ *vs.* $H_a\!: \delta \neq 0$ is

$$Z = \frac{\hat{\delta} - \delta}{\sqrt{Var\left(\hat{\delta}\right)}} \tag{18.5}$$

18.3.2 Random-Effects Meta-analysis Models

The fixed-effects MA model is simple, but the homogeneity assumption for between-study variation is often too restrictive. Intuitively, when we synthesize a group of studies with MA, one expects these studies to have enough in common to merit combination of the information for statistical inference. However, it is impractical to expect that all included studies have identical true effect size. Thus, the homogeneity assumption needs to be relaxed so practical applications may include situations where there is between-study variation in the true effect size of the studies. We need to allow for heterogeneity. The RE-MA model is used when the included studies are heterogeneous.

The random-effects model assumes that the underlying population effect sizes δ_k are normally distributed with a global mean of $'$ and a between-study variance of τ^2, [$\delta_k \sim N(\delta, \tau^2)$]. Therefore, the RE-MA is

$$\hat{\delta}_k = \delta_k + \varepsilon_k = \delta + \upsilon_k + \varepsilon_k \tag{18.6}$$

where $\upsilon_k \sim N(0, \tau^2)$. Moreover, the RE-MA model based on (18.6) is expressed in two levels:

$$\text{Level 1: } \hat{\delta}_k \sim N\left(\delta_k, \sigma_k^2\right) \tag{18.7}$$

$$\text{Level 2: } \delta_k \sim N(\delta, \tau^2) \tag{18.8}$$

The global population effect size for RE-MA is estimated using a weighted-mean methods, but the weights in RE-MA must incorporate both within-study variance (σ_k^2) and between-study variance (τ^2) as $w_{kR} = 1/(\sigma_k^2 + \hat{\tau}^2)$ where w_{kR} denotes weight for the kth study under a RE-MA model. Thus, an estimate of global effect size in RE-MA is

$$\hat{\delta}_R = \frac{\sum_{k=1}^{K} w_{\mathrm{kR}} \hat{\delta}_k}{\sum_{k=1}^{K} w_{\mathrm{kR}}} \tag{18.9}$$

with variance

$$\mathrm{Var}\left(\hat{\delta}_{\mathrm{R}}\right) = 1/\sum_{k=1}^{K} w_{\mathrm{kR}} \tag{18.10}$$

Such values lead to a 95% confidence interval for the global ES of δ in RE-MA as

$$\hat{\delta}_{\mathrm{R}} \pm 1.96\sqrt{\mathrm{Var}\left(\hat{\delta}_{\mathrm{R}}\right)} \tag{18.11}$$

and a test statistic for $H_0: \delta = 0$ vs. $H_a: \delta \neq 0$ as

$$Z = \frac{\hat{\delta}_R - \delta}{\sqrt{\mathrm{Var}\left(\hat{\delta}_R\right)}} \tag{18.12}$$

It is necessary to estimate the between-study variance τ^2 as it is needed in fitting the random effect meta-analysis model. These include the DerSimonian–Laird's method of moments ([MM] Dersimonian and Laird 1986), the maximum likelihood estimation (MLE) method (Hardy and Thompson 1998), the restricted maximum likelihood (REML) method (Raudenbush and Bryk 1985), and the Sidik–Jonkman (SJ) estimator (Sidik and Johkman 2005). The most commonly used estimators, MM and SJ, are distribution-free and are obtained based on a noniterative approach. The MLE and REML are parametric methods and need iterations for estimating τ^2. A comparison of these estimators is made in Sidik and Jonkman (2005). We review the most commonly used estimator, the DerSimonian–Laird MMs, as

$$\hat{\tau}^2 = \max\left(0, \frac{Q-(K-1)}{U}\right) \tag{18.13}$$

where U is a constant defined as $U = \sum_{k=1}^{K} \mathrm{w}_k - \frac{\sum_{k=1}^{K} w_k^2}{\sum_{k=1}^{K} w_k}$, and Q is the weighted sum of squared errors,

$$Q = \sum_{k=1}^{K} w_k \left(\hat{\delta}_k - \hat{\delta}\right)^2 \tag{18.14}$$

18.3.3 Testing and Quantifying the Between-Study Heterogeneity

Testing: The Q-statistic, Equation (18.14), is commonly used to test for heterogeneity across studies. This statistic follows as a chi-square (χ^2) distribution with $K - 1$ degrees of freedom. Under the null hypothesis of no heterogeneity, τ^2 is zero, and the expected value of Q is its degrees of freedom.

Quantification: While the Q-statistic is used to test for heterogeneity, it does not report on the extent of the heterogeneity. However, in order to quantify the heterogeneity, there are indices such as the τ^2-index, H-index, and I^2-index. These indices are based on the Q-statistic as described below:

τ^2-index: The τ^2*-index* is estimated using the Q-statistic:

$$\hat{\tau}^2 = \frac{Q-(K-1)}{U} \tag{18.16}$$

which is the DerSimonian–Laird MMs. Even though the true variance τ^2 can never be less than zero, the estimated variance $\hat{\tau}^2$ can be from the sampling error leading to $Q < K - 1$. When this occurs, the estimated $\hat{\tau}^2$ is set to zero.

The H-index (Higgins and Thompson 2002): The H-index, defined as $H = \sqrt{\frac{Q}{K-1}}$, is based on the fact that $E[Q] = K - 1$ when there is no heterogeneity. In this case, H is equal to 1. The H-index is greater than 1, as it measures the relative magnitude of heterogeneity among all studies. If the lower limit of this interval is greater than 1, the H is statistically significant.

The I^2*-index*: The I^2 index measures the proportion of observed heterogeneity from the real heterogeneity as $I^2 = \frac{Q-(K-1)}{Q} \times 100\%$, representing the ratio of excess dispersion to total dispersion. It is similar to the coefficient of determination in classical regression. It represents the proportion of the total variance explained by the regression variables. Higgins et al. (2003) suggested a value of the I^2-index around 25%, 50%, and 75% is considered as low-, moderate-, and high-heterogeneity, respectively.

18.4 Meta-regression

Meta-regression resembles a weighted regression model. The study-level moderators associated with the reported effect sizes serve as the dependent variable and their variances as weights. It is a typical multiple regression applied for study-level data. As such, the theory of regression is directly applicable to meta-regression.

While the RE-MA model in equation (18.6) takes into account between-study heterogeneity, it is not a tool to explain and explore how these heterogeneities are related to other study-level variables. However, when there are study-level variables (called "moderators"), the meta-regression is used to investigate the association between these moderators and the reported effect sizes.

There are two types of meta-regression: one based on the fixed-effects model and the other on the random-effects model. Suppose that there are p moderators $x_1, x_2, \ldots, x_p$ from K independent studies. The fixed-effects meta-regression model with regression parameters $\beta_0, \beta_1, \beta_2, \ldots, \beta_p$ is

$$\hat{\delta}_k = \beta_0 + \beta_1 x_{1k} + \beta_2 x_{2k} + \cdots + \beta_p x_{pk} + \varepsilon_k$$

This fixed-effects meta-regression model assumes that the variation in effect sizes can be completely explained by these moderators. In other words, the variation is predictable and hence the fixed-effects meta-regression.

The random-effects meta-regression can be obtained by adding random-effects b_k to the fixed-effects model above:

$$\hat{\delta}_k = \beta_0 + \beta_1 x_{1k} + \beta_2 x_{2k} + \cdots + \beta_p x_{pk} + b_k + \varepsilon_k$$

where b_k is assumed independent with a mean 0 and variance τ^2. Unlike the fixed-effects meta-regression, where all variability in effect sizes can be explained by the moderators $x_1, x_2, \ldots, x_p$, the random-effects meta-regression assumes that the model can explain only part of the variation, and random-effects b_k account for the remaining variation.

Meta-regression is most appropriate when a meta-analysis includes a large number of studies. Furthermore, since the moderators and outcome in meta-regression are all study-level summary statistics (e.g., patient mean age, proportion of female patients), the relation between these moderators and outcomes may not directly reflect the relation between subject scores and subjects' outcomes, causing aggregation bias. Therefore, careful consideration and interpretation of the results are always recommended when performing meta-regression, as discussed in Sutton et al. (2000).

18.5 Meta-analysis Using SAS

We use 13 clinical trials to evaluate the impact of the BCG vaccine on TB prevention, illustrating MA and meta-regression using SAS.

18.5.1 Meta-data and Effect Size

```
proc import out=BCG datafile=" "your data path"\BCG.xlsx"
dbms=xlsx replace;
Getnames=Yes;
run;

proc print data=BCG;run;
***Calculating Effect-size of log-OR;
data new (drop=pt pc);
set BCG;
pt=tpos/(tpos+tneg);
pc=cpos/(cpos+cneg);
Yi=log((pt/(1-pt))/(pc/(1-pc)));
Vi=1/tpos+1/cpos+1/(tneg-tpos)+1/(cneg-cpos);
run;
proc print data=new;
run;
```

This will produce the data as follows :

trial	author	year	tpos	tneg	cpos	cneg	ablat	y_i	v_i
1	Aronson	1948	4	119	11	128	44	-0.9387	0.3571
2	Ferguson & Simes	1949	6	300	29	274	55	-1.6662	0.2081
3	Rosenthal et al	1960	3	228	11	209	42	-1.3863	0.4334
4	Hart & Sutherland	1977	62	13536	248	12619	52	-1.4564	0.0203
5	Frimodt-Moller et al	1973	33	5036	47	5761	13	-0.2191	0.0520

trial	author	year	tpos	tneg	cpos	cneg	ablat	y_i	v_i
6	Stein & Aronson	1953	180	1361	372	1079	44	-0.9581	0.0099
7	Vandiviere et al	1973	8	2537	10	619	19	-1.6338	0.2270
8	TPT Madras	1980	505	87886	499	87892	13	0.0120	0.0040
9	Coetzee& Berjak	1968	29	7470	45	7232	27	-0.4717	0.0570
10	Rosenthal et al	1961	17	1699	65	1600	42	-1.4012	0.0754
11	Comstock et al	1974	186	50448	141	27197	18	-0.3408	0.0125
12	Comstock& Webster	1969	5	2493	3	2338	33	0.4466	0.5342
13	Comstock et al	1976	27	16886	29	17825	33	-0.0173	0.0716

Two additional variables (y_i and v_i) included are the calculated effect-size (i.e. log-OR in this case) and their associated variances. An examination of these data show that 11 of 13 studies (except # 8 and #12) have negative log-ORs, which indicate that the BCG vaccine reduced the risk of TB for these 11 studies, but not for the trials of 8 and 12.

18.5.2 Fixed-Effects Meta-analysis Models

```
proc mixed data=new method=ml;
class trial;
model Yi=/s cl;
repeated / group=trial;
parms
(0.35815)(0.20863)(0.43374)(0.02032)(0.05195)
(0.0105)(0.22704)(0.00401)(0.05698)(0.07545)(0.01253)
(0.53416)(0.07164)/eqcons=1 to 13;
run;
```

The mixed procedure output is as follows:

Model Information	
Data Set	WORK.NEW
Dependent Variable	Yi
Covariance Structure	Variance Components
Group Effect	Trial
Estimation Method	ML

Class Level Information

Class	Levels	Values
trial	13	1 2 3 4 5 6 7 8 9 10 11 12 13

Covariance Parameter Estimates

Cov Parm	Group	Estimate	
Residual	trial 1	0.3582	
Residual	trial 2	0.2086	
Residual	trial 3	0.4337	
Residual	trial 4	0.02032	
Residual	trial 5	0.05195	
Residual	trial 6	0.01050	
Residual	trial 7	0.2270	
Residual	trial 8	0.004010	
Residual	trial 9	0.05698	
Residual	trial 10	0.07545	
Residual	trial 11	0.01253	
Residual	trial 12	0.5342	
Residual	trial 13	0.07164	
-2 Log Likelihood			150.5
AIC (Smaller is Better)			152.5
AICC (Smaller is Better)			152.8
BIC (Smaller is Better)			153.0

Solution for Fixed-effects

Effect	Estimate	Standard Error	DF	t Value	Pr>|t|	Alpha	Lower	Upper
Intercept	-0.4308	0.04249	12	-10.14	<.0001	0.05	-0.5234	-0.3383

Effect	Estimate	Standard Error	DF	t Value	Pr>|t|	Alpha	Lower	Upper
Intercept	-0.4308	0.04249	12	-10.14	<.0001	0.05	-0.5234	-0.3383

18.5.3 Random-Effects Meta-analysis Models

```
proc mixed data=new  method=ml;
class trial;
model Yi=/s cl;
repeated / group=trial;
```

```
random intercept/ subject=trial;
parms
(0)(0.35815)(0.20863)(0.43374)(0.02032)(0.05195)
(0.0105)(0.22704)(0.00401)(0.05698)(0.07545)(0.01253)
(0.53416)(0.07164)/eqcons=2 to 14;
run;
```

The mixed procedure output is as follows:

Model Information	
Data Set	WORK.NEW
Dependent Variable	Yi
Covariance Structure	Variance Components
Subject Effect	trial
Group Effect	trial
Estimation Method	ML
Residual Variance Method	None
Fixed-effects SE Method	Model-Based
Degrees of Freedom Method	Containment

Class Level Information

Class	Levels	Values
trial	13	1 2 3 4 5 6 7 8 9 10 11 12 13

Dimensions		
Covariance Parameters	14	
Columns in X	1	
Columns in Z per Subject	1	
Subjects	13	
Max Obs per Subject	1	
Number of Observations		
Number of Observations Read		13
Number of Observations Used		13
Number of Observations Not Used		0

Iteration History

Iteration	Evaluations	-2 Log Like	Criterion
1	2	112.78836245	0.46295625
2	1	85.09580282	0.49343958
3	1	64.80309725	0.53199459
4	1	50.21729236	0.57781371
5	1	40.06465075	0.62618654

Iteration History			
Iteration	**Evaluations**	**-2 Log Like**	**Criterion**
6	1	33.35167676	0.65554252
7	1	29.28958038	0.60800481
8	1	27.18351550	0.40658743
9	1	26.35558814	0.13436910
10	1	26.16368968	0.01282513
11	1	26.14823255	0.00012726
12	1	26.14808804	0.00000001
13	1	26.14808802	0.00000000

```
Convergence criteria met.
```

Convergence is met so we can accept these parameter estimates. There are 14 covariance parameters in the Random-effects model. The 13 studies plus intercept.

Covariance Parameter Estimates			
Cov Parm	**Subject**	**Group**	**Estimate**
Intercept	trial		0.3025
Residual		trial 1	0.3582
Residual		trial 2	0.2086
Residual		trial 3	0.4337
Residual		trial 4	0.02032
Residual		trial 5	0.05195
Residual		trial 6	0.01050
Residual		trial 7	0.2270
Residual		trial 8	0.004010
Residual		trial 9	0.05698
Residual		trial 10	0.07545
Residual		trial 11	0.01253
Residual		trial 12	0.5342
Residual		trial 13	0.07164

Fit Statistics		
-2 Log Likelihood		26.1
AIC (Smaller is Better)		30.1
AICC (Smaller is Better)		31.3
BIC (Smaller is Better)		31.3
PARMS Model Likelihood Ratio Test		
DF	Chi-Square	Pr > ChiSq
1	124.32	<.0001

The likelihood ratio test has a value of 124.32 with significant p-value <.0001.

Solution for Random-effects								
Effect	**Estimate**	**Standard Error**	**DF**	**t Value**	**Pr>|t|**	**Alpha**	**Lower**	**Upper**
Intercept	-0.7418	0.1780	12	-4.17	0.0013	0.05	-1.1296	-0.3540

The global overall estimate of log-OR from the random-effects meta-analysis model is -0.7418 with SE = 0.1780 and with p-value < 0.0013, which indicates statistical significance if meta-analyzed together.

18.5.4 Meta-regression

```
proc mixed data=new  method=ml CONVH=1E-7;
class trial;
model Yi=ablat/s cl;
repeated / group=trial;
random intercept/ subject=trial;
parms
(0)(0.35815)(0.20863)(0.43374)(0.02032)(0.05195)
(0.0105)(0.22704)(0.00401)(0.05698)(0.07545)
(0.01253) (0.53416)(0.07164)/eqcons=2 to 14;
run;
```

The mixed procedure output is as follows:

Model Information	
Data Set	WORK.NEW
Dependent Variable	Yi
Covariance Structure	Variance Components
Subject Effect	trial
Group Effect	trial
Estimation Method	ML
Residual Variance Method	None
Fixed-effects SE Method	Model-Based
Degrees of Freedom Method	Containment

Class Level Information		
Class	**Levels**	**Values**
trial	13	1 2 3 4 5 6 7 8 9 10 11 12 13

Dimensions	
Covariance Parameters	14
Columns in X	2
Columns in Z per Subject	1
Subjects	13
Max Obs per Subject	1

Iteration History			
Iteration	**Evaluations**	**-2 Log Like**	**Criterion**
1	2	13.95331826	0.00444385
2	2	13.93645138	0.00018991
3	1	13.93547541	0.00000049
4	1	13.93547294	0.00000000

```
Convergence criteria met.
```

Covariance Parameter Estimates			
Cov Parm	**Subject**	**Group**	**Estimate**
Intercept	trial		0.003762
Residual		trial 1	0.3582
Residual		trial 2	0.2086
Residual		trial 3	0.4337
Residual		trial 4	0.02032
Residual		trial 5	0.05195
Residual		trial 6	0.01050
Residual		trial 7	0.2270
Residual		trial 8	0.004010
Residual		trial 9	0.05698
Residual		trial 10	0.07545
Residual		trial 11	0.01253
Residual		trial 12	0.5342
Residual		trial 13	0.07164

Estimates of the covariance parameters for each trial is given in the last column. They measure the variation for each trial.

Fit Statistics	
-2 Log Likelihood	13.9
AIC (Smaller is Better)	19.9
AICC (Smaller is Better)	22.6
BIC (Smaller is Better)	21.6

PARMS Model Likelihood Ratio Test

DF	Chi-Square	Pr > ChiSq
1	0.05	0.8198

Solution for Fixed-effects

Effect	Estimate	Standard Error	DF	t Value	Pr>|t|	Alpha	Lower	Upper
Intercept	0.3729	0.1049	11	3.55	0.0045	0.05	0.1420	0.6039
ablat	-0.03280	0.003360	0	-9.76	.	0.05	.	.

Meta-regression summary on log-OR to the absolute latitude with estimated intercept of -0.3729 and slope of -0.03280.

18.6 Meta-analysis Using R

18.6.1 Data and R Package "Metafor"

There are several R packages for MA, such as "meta", "rmeta" and "metafor" (Chen and Peace 2013). We recommend the R package "metafor" for its functionalities and ease of implementation. The "metafor" package requires all the metadata to be converted to the effect size or treatment effect between groups. The MA combines the individual estimates of ESs across studies. If the effect sizes is not provided for the individual study, but the number of patients who respond to treatment is provided (as seen in Table 18.1), then one needs to calculate the effect size for each study and subsequently pool the estimated effect sizes across all of the studies in a MA. The calculation is done with function "escalc" in "metafor" package. This is used to convert any data type of metadata to effect sizes. We illustrate the use of binomial data in Table 18.1.

Let's load the R package and the data into R based on R code chunk:

```
># load the R library
> library("metafor")
># load the BCG data
> data("dat.bcg", package = "metafor")
># Print the data
> dat.bcg
```

trial	author	year	tpos	tneg	cpos	cneg	ablat
1	Aronson	1948	4	119	11	128	44
2	Ferguson & Simes	1949	6	300	29	274	55
3	Rosenthal et al	1960	3	228	11	209	42
4	Hart & Sutherland	1977	62	13536	248	12619	52
5	Frimodt-Moller et al	1973	33	5036	47	5761	13
6	Stein & Aronson	1953	180	1361	372	1079	44
7	Vandiviere et al	1973	8	2537	10	619	19
8	TPT Madras	1980	505	87886	499	87892	13
9	Coetzee& Berjak	1968	29	7470	45	7232	27
10	Rosenthal et al	1961	17	1699	65	1600	42
11	Comstock et al	1974	186	50448	141	27197	18
12	Comstock & Webster	1969	5	2493	3	2338	33
13	Comstock et al	1976	27	16886	29	17825	33

18.6.2 Defining Effect Size

The data in Table 18.1 follow a binomial distribution, illustrating the number of TB-positive (TB+) and TB-negative (TB–) cases from both BCG vaccine and control groups. Consider the probability (we call it "risk" here) of TB+ and compare the probabilities (i.e., risks) between groups to measure the efficacy of BCG vaccine. The most commonly used effect size is the risk difference (RD), which is simply the difference of the risks between a treatment (T, for example, the BCG vaccine in this data) group and control (C) group, defined as $\text{RD} = p_T - p_C$ with $p_T = \text{tpos}/(\text{tpos} + \text{tneg})$ and $p_C = \text{cpos}/(\text{cpos} + \text{cneg})$ corresponding to the data from Table 18.1. The statistical inference for RD is to test whether the difference between this RD and zero is statistically significant.

The second, effect size is the risk ratio, which is defined as the ratio of the risk from the treatment to control: $\text{RR} = \frac{p_T}{p_C}$, where the statistical inference for RR is to test whether the difference between this RR and 1 is statistically significant (Chen and Peace 2013).

The third is commonly used effect size, the odds ratio (OR) which is originated from the logistic regression to analyze the binomial data. The OR associated with an event is defined as the ratio of the odds of the event in one study group to the odds of the event in another study group. The odds of the event is defined as $\text{odds} = \frac{p}{1-p}$. Thus, the OR of the treatment group (such as the BCG vaccine) to the control group for each study is $\text{OR} = \frac{\text{Odds}_T}{\text{Odds}_C} = \frac{\frac{p_T}{1-p_T}}{\frac{p_C}{1-p_C}}$, where the statistical inference for OR is to test whether the difference between this OR and one is statistically significant. Specifically, this OR is transformed to log OR so the log-OR is approximately normally distributed.

18.6.3 Calculating Effect Size

For illustration, we use effect size of log-OR as in following R code chunk.

```
> Dat = escalc(measure="OR",ai=tpos,bi=tneg,ci=cpos,di=cneg,
data = dat.bcg, append = TRUE)
```

The option append=TRUE is to append the calculated ES, i.e. log-OR, to the original data. We can print the new data frame as follows:

```
# print the data
>Print(dat)
```

trial	author	year	tpos	tneg	cpos	cneg	ablat	y_i	v_i
1	Aronson	1948	4	119	11	128	44	-0.9387	0.3571
2	Ferguson & Simes	1949	6	300	29	274	55	-1.6662	0.2081
3	Rosenthal et al	1960	3	228	11	209	42	-1.3863	0.4334
4	Hart & Sutherland	1977	62	13536	248	12619	52	-1.4564	0.0203
5	Frimodt-Moller et al	1973	33	5036	47	5761	13	-0.2191	0.0520
6	Stein & Aronson	1953	180	1361	372	1079	44	-0.9581	0.0099
7	Vandiviere et al	1973	8	2537	10	619	19	-1.6338	0.2270
8	TPT Madras	1980	505	87886	499	87892	13	0.0120	0.0040
9	Coetzee& Berjak	1968	29	7470	45	7232	27	-0.4717	0.0570
10	Rosenthal et al	1961	17	1699	65	1600	42	-1.4012	0.0754
11	Comstock et al	1974	186	50448	141	27197	18	-0.3408	0.0125
12	Comstock & Webster	1969	5	2493	3	2338	33	0.4466	0.5342
13	Comstock et al	1976	27	16886	29	17825	33	-0.0173	0.0716

In the new data frame 'dat", two additional variables (y_i and v_i) included are the calculated effect size (i.e., log-OR in this case) and their associated variances. An examination of these data shows that 11 of 13 studies (except # 8 and #12) have negative log-ORs, which indicate that the BCG vaccine reduced the risk of TB for these 11 studies but not for the trials of 8 and 12. Their statistical significance can be examined by constructing the *z*-statistic using the ratio of y_i to the squared-root of v_i or the 95% confidence intervals, as shown in Figure 18.1. Examining the top panel in this figure, we can see that only studies 2, 3, 4, 6, 7, 10 and 11 (i.e., 7 out of 13) are statistically significant.

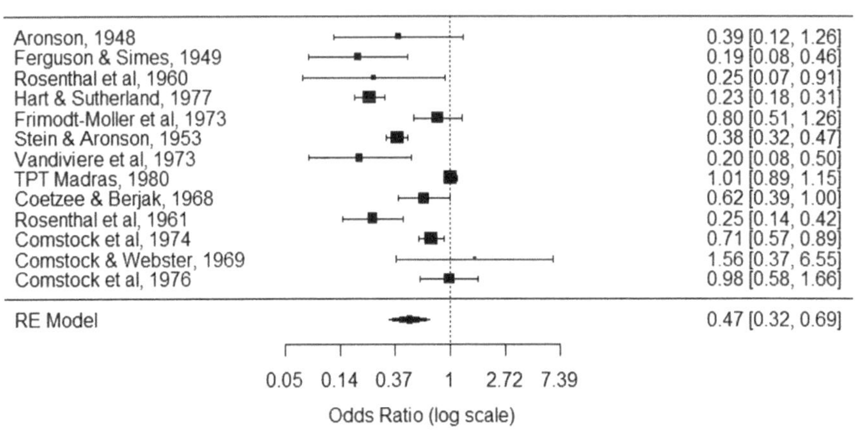

FIGURE 18.1
Data and 95% Confidence Intervals of ORs for 13 Studies along with the RE-MA.

18.6.4 Fixed-Effects Meta-analysis

A MA of the 13 studies is conducted under the assumption of homogeneity for 13 studies. The fixed-effects MA with R function "rma" is

```
# Call 'rma' to fit the BCG data with Fixed-effect MA
>meta.FE = rma(yi,vi,data=dat,method="FE",
          slab=paste(author,year,sep=", "))
# Print the summary
> meta.FE
```

This produces R output for the fixed-effects MA as follows:

```
Fixed-Effects Model (k = 13)
Test for Heterogeneity:
Q(df = 12) = 163.1649, p-val < .0001
Model Results:
estimate     se      zval     pval    ci.lb     ci.ub
-0.4361  0.0423 -10.3190  <.0001  -0.5190   -0.3533    ***
```

The test of heterogeneity as shown with Q-statistic: Q = 163.1649, degrees of freedom 12 with p-value < 0.0001 from the χ^2-test shows as highly statistically significant. It indicates the assumption of homogeneous studies in fixed-effects meta-analysis is not appropriate. Thus, random-effects meta-analysis is considered as follows.

The estimated log-OR from combining the 13 studies is -0.4361 with SE of 0.0423, zval of -10.3190 and p-value < 0.0001. The associated 95% confidence interval for the global log-OR is (-0.5190, -0.3533). This means that in a meta-analysis of these 13 studies, the global overall effect of BCG vaccine is statistically significant in preventing TB.

18.6.5 Random-Effects Meta-analysis

The Dersimonian–Laird RE-MA model is implemented as

```
# Call 'rma' with method="DL" to fit the random-effects MA
>meta.DL=rma(yi,vi,data=dat,method="DL",
                              slab=paste(author,year,sep=", ")))
# Print the summary
> meta.DL
```

This produces the output for random-effects MA as follows:

```
Random-Effects Model (k = 13; tau^2 estimator: DL)
tau^2        (estimated        amount        of        total
heterogeneity): 0.3663 (SE = 0.2659)
tau (square root of estimated tau^2 value):    0.6053
I^2 (total heterogeneity / total variability):  92.65%
H^2 (total variability / sampling variability):  13.60
Test for Heterogeneity:
Q(df = 12) = 163.1649, p-val < .0001
Model Results:
estimate      se      zval    pval    ci.lb    ci.ub
-0.7474  0.1923  -3.8873  0.0001  -1.1242  -0.3706    ***
```

The global overall estimate of log-OR from the random-effects meta-analysis model is -0.7474 with SE = 0.1923 with p-value < 0.0001, which indicates statistical significance if meta-analyzed together. This RE-MA is graphically illustrated with the forest plot in a simple R command as follows:

```
># Forest plot for the RE-MA with "exp" transformation to show ORs
>forest(meta.DL, atransf=exp)
```

This forest plot (see Figure 18.1) shows the study-specific ES along with the 95% confidence intervals. The bottom panel is the MA for all 13 studies. Further examination of the R output in the random-effects meta-analysis, the estimated between-study heterogeneity $\hat{\tau}^2 = 0.3663$ with $I^2 = 92.65\%$ and $H^2 = 13.60$, which demonstrates highly and statistically significant heterogeneity. Meta-regression should be used to explain this heterogeneity.

18.6.6 Meta-regression

The between-study heterogeneity is explained with the use of the geographical latitude where the clinical trials were conducted. This is introduced as random effects. This random-effects meta-regression is modeled R code chunk:

```
> # Meta-regression
> metaReg.ablat = rma(yi,vi,mods = ~ablat, data = dat,method="DL",
   slab=paste(author, year, sep=", "))
> # Print the meta-regression results
     > metaReg.ablat
Mixed-Effects Model (k = 13; tau^2 estimator: DL)
tau^2 (estimated amount of residual heterogeneity):
                                          0.0480 (SE = 0.0451)
tau (square root of estimated tau^2 value):              0.2191
I^2 (residual heterogeneity /
unaccounted variability): 56.17%
H^2 (unaccounted variability / sampling variability):  2.28
R^2 (amount of heterogeneity accounted for):            86.90%
Test for Residual Heterogeneity:
QE(df = 11) = 25.0954, p-val = 0.0088
Test of Moderators (coefficient(s) 2):
QM(df = 1) = 26.1628, p-val < .0001
Model Results:
estimate       se      zval    pval    ci.lb    ci.ub
intrcpt    0.3030 0.2109  1.4370  0.1507   -0.1103  0.7163
ablat     -0.0316 0.0062 -5.1150 <.0001  -0.0437  -0.0195  ***
```

The study-level moderator, ablat, the estimated residual heterogeneity $\hat{\tau}^2$ reduced to 0.0480 from the previous random-effect meta-analysis without ablat where the heterogeneity was estimated at 0.3663. This indicates that the moderator ablat itself accounts for (0.3663-0.0480)/

0.3663 = 87% of the total amount of heterogeneity, and the absolute latitude is significantly related to the effectiveness of the BCG vaccine in preventing TB, which is quantified in the estimated meta-regression equation:

$$\log(OR) = 0.3030\text{-}0.0316\,\textit{ablat}\,.$$

This estimated meta-regression equation and the entire meta-regression summary are graphically displayed in Figure 18.2.

This figure is produced in the following R code chunk with detailed explanations following "#":

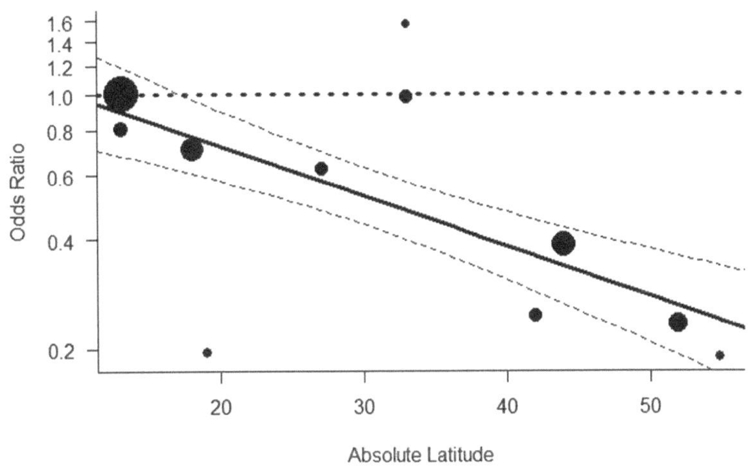

FIGURE 18.2
Results from Meta-regression.

```
# Create a new latitude vector from 0 to 60 for smooth plotting
>newlat = 0:60
# Using the meta-regression to calculate the predicted ORs
>Pred.OR  = predict(metaReg.ablat, newmods = n ewlat, transf = exp)
# Use the inverse-variance to create a weighting for the data
>wi     = 1/sqrt(dat$vi)
>size   = 1 + 3 * (wi - min(wi))/(max(wi) - min(wi))
# Scatter Plot the ORs
>plot(dat$ablat, exp(dat$yi), pch = 19, cex = size,
xlab = "Absolute Latitude", ylab = "Odds Ratio",
las  = 1, bty = "l", log = "y")
# Add a thicker line for the meta-regression using "lwd=3"
>lines(newlat, Pred.OR$pred, lwd=3)
# Add the CIs
>lines(newlat, Pred.OR$ci.lb, lty = "dashed")
>lines(newlat, Pred.OR$ci.ub, lty = "dashed")
# Add a dotted horizontal line for equal-effectiveness
>abline(h = 1, lwd=3,lty = "dotted")
```

In this figure, the meta-regression is plotted in the solid line, and its 95% CI bounds are plotted in dashed dots. The horizontal dotted line from OR = 1 indicates no BCG vaccine effects. The 13 studies are plotted in 13 solid dots, and the sizes of the dots are proportional to their calculated variance.

It can be seen from this equation and the associated Figure 18.2 that the higher the absolute latitude, the more effective is the BCG vaccine. When the ablat is less than 20 degrees and close to zero (i.e., study performed closer to the equator), the OR would be close to zero (as evidenced from the insignificant intercept parameter and Figure 18.2, which means that the vaccination has no real effect on prevention TB. As ablat increases, for example, to a latitude of 60 degrees, the log-OR as calculated from the estimated meta-regression equation log(OR) = 0.3030-0.0316 *ablat* is −1.593, which corresponds to an OR of 0.20.

We redraw the forest plot in Figure 18.1 after incorporating the meta-regression with the moderator "ablat", as shown in the following R code chunk, which produces Figure 18.3:

```
### Forest plot after meta-regression with ablat annotated
>forest(metaReg.ablat, xlim=c(-9,5), ilab=dat.bcg$ablat,
     ilab.xpos=-3.5, atransf=exp, at=log(c(.05,.25,1,4)))
```

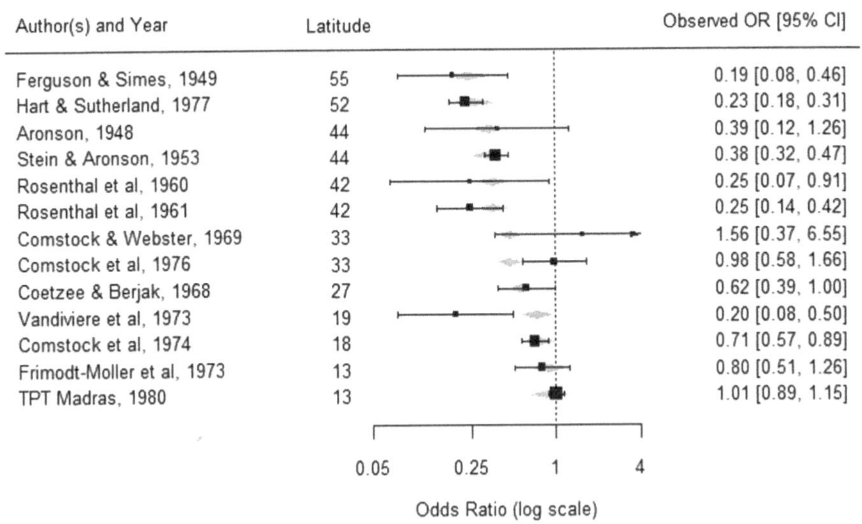

FIGURE 18.3
Forest Plot after Meta-regression.

```
# Add texts to the plot
>text(-9,  15, "Author(s) and Year",    pos=4, cex=1.1)
>text( 5,  15, "Observed OR [95% CI]",  pos=2, cex=1.1)
>text(-3.5, 15, "Latitude",                    cex=1.1)
```

In this figure, the right panel is to show the study-specific point estimates of OR and their 95% CIs for each of these 13 studies. The middle panel is the graphical representation of these 95% CIs in solid line segments. The shaded diamonds are the 95% CIs after the meta-regression adjustment. At the left side of these line segments of CIs, we show the corresponding latitudes.

18.7 Summary and Discussions

This chapter provides an overview of MA using classical fixed-effects and random-effects models, illustrating their application with SAS "proc mixed" and the "metafor" package in R. Other commonly used R packages for conducting MA, such as "rmeta" and "meta", and extensive illustrations of their applications can be freely accessed from www.r-project.org. For further details on MA methodology and its implementation in R, see *Applied Meta-Analysis with R* by Chen and Peace (2013).

18.8 Exercises

In this chapter, we illustrated the RE-MA using the classical DerSimonian–Laird moment estimator as specified in R option *method*="DL" to estimate the between-study heterogeneity. There are several other methods to estimate the between-study heterogeneity. This exercise aims to familiarize readers with alternative estimators, such as:

- method="HE" = Hedges estimator
- method="HS" = Hunter-Schmidt estimator
- method="SJ" = Sidik–Jonkman estimator
- method="ML" = maximum-likelihood estimator
- method="REML" = restricted maximum-likelihood estimator
- method="EB" = empirical Bayes estimator
- method="PM" = Paule-Mandel estimator
- method="GENQ" = generalized Q-statistic estimator

Reanalyze the data using the listed methods and compare the resulting meta-analyses.

References

Borenstein, M., Hedges, L. V., Higgins, J. P. T., and Rothstein, H. R. (2009). *Introduction to Meta-Analysis*. Chichester, UK: Wiley.

Chen, D. G. and Peace, K. E. (2013). *Applied Meta-Analysis with R*. Boca Raton, FL: Chapman and Hall/CRC Biostatistics Series.

DerSimonian, R. and Laird, N. (1986). Meta-analysis in clinical trials. *Controlled Clinical Trials*, *7*, 177–188.

Hardy, R. J., & Thompson, S. G. (1998). Detecting and describing heterogeneity in meta-analysis. *Statistics in Medicine*, *17*(8), 841–856.

Hartung, J., Knapp, G., and Sinha, B. K. (2008). *Statistical Meta-Analysis with Applications*. Hoboken: Wiley.

Higgins, J. P. T., & Thompson, S. G. (2002). Quantifying heterogeneity in a meta-analysis. *Statistics in Medicine*, *21*(11), 1539–1558.

Higgins, J. P. T., Thompson, S. G., Deeks, J. J., & Altman, D. G. (2003). Measuring inconsistency in meta-analyses. *BMJ*, *327*(7414), 557–560.
Pigott, T. D. (2012). *Advances in Meta-Analysis*. New York: Springer.
Raudenbush, S. W. and Bryk, A. S. (1985). Empirical Bayes meta-analysis. *Journal of Educational Statistics*, *10*(2), 75–98.
Schwarzer, G., Carpenter, J. R., and Rucker, G. (2015). *Meta-analysis with R*. New York: Springer.
Sidik, K. and Jonkman, J. N. (2005). Simple heterogeneity variance estimation for meta-analysis. *Journal of Royal Statistical Society, Series C Applied Statistics*, *54*, 367–384.
Sutton, A. J., Abrams, K. R., Jones, D. R., Sheldon, T. A., & Song, F. (2000). *Methods for Meta-Analysis in Medical Research*. Chichester, UK: John Wiley & Sons.
Whitehead, A. (2003). *Meta-Analysis of Controlled Clinical Trials*. New York: John Wiley.

19

Spatial Statistical Analysis

19.1 Research Interest/Questions

In most research and public health applications, data are often obtained from spatial survey sampling. Spatial dependence is a fundamental characteristic of such data. It is intuitive that data at proximal locations would be correlated to each other with a higher correlation at closer proximities. This spatial dependency leads to spatial autocorrelation, which is different from the classical statistical modeling where independence among observations is assumed. For example, the regression models discussed in Chapters 4 to 9 require that the observations in the data are independent and do not compensate for spatial dependency. These classical regression models, when applied to spatial data, result in unstable parameter estimates and unreliable significance tests. Public health recommendations and decision-making based on these classical statistical models would be misleading.

To address this phenomenon, spatial autocorrelation should be incorporated in the analysis, which is the principle in spatial statistical analysis. Spatial statistics extend these classical models and analyze the spatially autocorrelated data. In particular, it is often the case that observations on significant covariates are not available, leaving only the detection of spatial autocorrelation in the data, or that model residuals are the only way to model the remaining variation.

Extensive research and applications have been conducted on spatial statistics. To name a few, Cressie (1991, 2015) and Sudipto et al. (2014) provide a comprehensive background on the theoretical development of spatial statistics. A specific Elsevier journal, "Spatial Statistics", is dedicated to the development and applications in spatial statistics (www.journals.elsevier.com/spatial-statistics). There are also countless online documentation and discussion groups devoted to spatial statistics.

This chapter gives an introductory overview of spatial statistics with implementation in commonly used statistical software such as SAS and

DOI: 10.1201/9781003596370-8

R. We illustrate spatial analysis using the New York Leukemia dataset and the Meuse River dataset.

19.2 New York Leukemia Data and Meuse River Data

In many public health studies, we would like to use specific covariates to explore associations with the spatially observed outcome of interest and adjust our data maps for potentially important demographic risk factors.

In this chapter, a regression analysis (with spatially correlated errors) of the New York leukemia data is analyzed. We consider the census tract-level leukemia incidence data, covering an eight-county region of upstate New York. As an exposure covariate, consider the inverse distance between each census tract centroid to the nearest of the inactive hazardous waste sites containing trichloroethylene (TCE).

Using distance as a surrogate for exposure relies on several assumptions. In particular, we assume that exposure decreases monotonically with the distance from each waste site. Also, we assume that all people residing within a given census tract receive the same exposure and ignore the amount of TCE reported at each site, treating all sites as sources of equivalent exposure. Finally, we assume that the inverse distance to the *nearest* site represents the only relevant source of TCE exposure. The authors of these data declared that they attempted to assess the association between leukemia incidence and proximity to locations of sites reporting TCE and not to assess the association between leukemia incidence and TCE exposure itself. However, individual exposure is difficult and costly to measure; therefore, these assumptions allow us a preliminary indication of elevated leukemia risk that may be associated TCE exposure.

A subset of these data is presented in Table 19.1A. *Let* x and y denote locations of each census tract centroid. "exposure" denotes the inverse distance from the tract centroid to the nearest TCE site, and "P_own home" and "P_over 65" denote census variables on percentages of the home owner and residents over 65 years old (1980 U.S. Census).

We also give an example of using R through the Meuse River data. As described in the R package "sp", this dataset includes the 155 spatial locations along with their topsoil heavy metal concentrations. It also includes several soil and landscape variables at these sampling locations in a flood plain of the river Meuse (therefore denoted by data "meuse"), near the village of Stein, The Netherlands. The heavy metal concentrations are collected from composite samples of an area of approximately 15 m × 15 m.

Fourteen variables are in this dataset including: x = spatial Easting (m) in Rijksdriehoek (RDH) (Netherlands topographical) map coordinates;

TABLE 19.1A

A Subset of Observations in "Leukemia" Dataset

Obs	x	y	Exposure	P_OwnHome	P_Over65	Pop	Cases
1	−55.482	9.899	0.059	0.591	0.116	2919	3.511
2	−54.284	36.22	0.071	0.776	0.102	3905	1.684
3	−53.546	21.137	0.13	0.721	0.159	5616	1.984
4	−51.984	50.333	0.038	0.684	0.103	2235	0.392
5	−51.912	−24.866	0.086	0.705	0.121	7041	2.081
6	−51.202	−0.781	0.04	0.558	0.119	1762	0.309
7	−51.142	−42.1	0.069	0.747	0.107	4401	1.051
8	−50.243	36.814	0.077	0.537	0.032	10494	3.839
9	−48.719	35.411	0.089	0.799	0.127	6501	1.139
10	−48.713	54.252	0.033	0.759	0.091	3571	0.626

y = spatial Northing (m) in RDH coordinates; cadmium = the topsoil cadmium concentration ("ppm"); copper = topsoil copper concentration ("ppm"); lead = topsoil lead concentration ("ppm"); zinc = topsoil zinc concentration ("ppm"); elev = relative elevation above local river bed (m); dist = distance to the Meuse river bank; om = organic matter (kg); ffreq = flooding frequency class; soil = soil type according to the 1:50 000 soil map of the Netherlands; lime = lime class; landuse = landuse class; and dist.m = distance to river Meuse in meters, as obtained during the field survey. A subset of these data is given in Table 19.1B.

The zinc (i.e., "zinc") concentration was used as an example in Bivand et al. (2008) to illustrate the R package "sp". In this chapter, we make use of the lead (i.e., "lead") concentration to illustrate spatial analysis. Figure 19.1 displays the lead concentration in the Meuse River. The closer to the riverbank, the larger the lead concentration, which illustrates the spatial trend. There is evidence that there are some spatial clustering in the data, which illustrate spatial correlations in the data. Therefore, spatial analysis is necessary for this type of data.

19.3 Models in Spatial Statistical Analysis

It is fully established that an understanding of location and place is an important element in effective decision-making. Advances in geospatial analytics have led to the transformation of science, society, and decision-making. Geospatial data (are also referred to as location data or spatial data) enable researchers to add the context of time and location to traditional data, so they can see changes over time as well as the locations of those changes.

TABLE 19.1B

A Subset of Observations in "Meuse" Dataset

X	*Y*	Cadmium	Copper	Lead	Zinc	Cellew	Dist	Om	Ffreq	Soil	Lime	Landuse	Callissttirm
181072	333611	11.7	85	299	1022	7.909	0.001	13.6	1	1	1.	Ah	50
181025	33358	8.6	81	277	1141	6.983	0.012	14	1	1	1	Ah	30
181165	333537	6.5	68	199	640	7.8	0.103	13	1	1	1	Ah	150
181298	333484	2.6	81	116	257	7.655	0.190	8	1	2	0	Ga	270
1811307	333330	2.8	48	117	269	7.48	0.277	8.7	1	2	0	Ah	380
181390	333260	3	61	137	281	7.791	0.364	7.8	1	2	0	Ga	470
181165	333370	3.2	31	132	346	8.217	0.190	9.2	1	2	0	Ah	240
181027	333363	2.8	29	150	406	8.49	0.092	9.5	1	1	0	Ab	120
181060	333231	2.4	37	133	347	8.668	0.185	10.6	1	1	0	Ab	240

FIGURE 19.1
Lead Concentration along the Meuse River.

Geospatial data provide answers not found in charts or tables. Examples are as follows:

1. What are customers buying, and where are they buying it?
2. What infectious diseases are on the rise?
3. Where are high unemployment or crime rates?
4. Who are your competitors, and where are they located?

Geospatial data integrate time and space elements, enhancing predictive models by incorporating temporal and spatial context. It identifies the geographic location of features, such as natural or constructed features, oceans, and more. Geospatial data often describes address-related data, which is matched to latitude and longitude using a process known as geocoding. The data are often analyzed through geographic information systems (GIS) along with advanced models in spatial statistics.

Geospatial data are usually gathered through satellite and ground-based radio navigation systems. A healthcare company can study disease outbreaks over time at specific locations and more adequately prepare for potential epidemics. Geospatial analytics helps with predictive insights by accounting for both time and space in predictive models. It allows one to more accurately forecasted events at a specific location for any future point in time.

The behavior of humans in many ways is related to the environment in which the behavior occurs. Often researchers ignored environment variables and focused on individual-level variables. These techniques presented make it possible for researchers to routinely address questions such as "What is the influence of neighborhood characteristics on student achievement"? This assumption is inconsistent with the first law of geography, mainly that objects closer together in space and time are more related than objects further away in space and time (Xu and Kennedy 2015). If this relationship exists and is ignored, then the standard errors are inaccurate. Spatial analysis techniques provide a means for testing the assumption by exploring issues of proximity (Haining 1990).

This chapter describes a number of techniques of spatial analysis and their potential for applications in public health epidemiology. We use SAS and R in our demonstration.

19.3.1 What Is Spatial Analysis?

Spatial analysis is an analysis of data that takes into consideration the location or distance between objects. It includes techniques to determine if data exhibit spatial autocorrelation and modeling any spatial relationships (Anselin 1995; Fotheringham et al. 2000). In health studies, spatial analysis, as well as maps of the spatial distribution of phenomena, such as epidemic level, can be useful to address the outbreak. We follow Xu and Kennedy (2015) and present several of the more popular spatial modeling approaches:

1. Does spatial autocorrelation exist?
2. Present techniques for modeling spatial relationships.
3. Provide several examples using the SAS and R software.

19.3.2 The Goal of Spatial Analysis

The goal of spatial data analysis is to derive insights from the location and context of real-world phenomena. Spatial analysts are often concerned with why things happen where they do. They are also concerned with how the occurrence of a phenomenon in a particular region affects nearby regions.

19.4 Spatial Analysis of Leukemia Data and Meuse River Data in SAS/R

In many public health studies, we use specific covariates to explore associations with the outcome of interest and adjust our maps for potentially important demographic risk factors. We turn now to linear regression analysis (with spatially correlated errors) of the New York leukemia data. We illustrate four regression models, which include two ordinary regression models with and without spatial autocorrelations and two weighted regression models with and without spatial autocorrelation.

19.4.1 Spatial Analysis of Leukemia Data Using SAS

```
data one;
set nydata;
logpop=log(pop8);
run;
proc sort;
by areakey;
run;
```

19.4.1.1 Model 1: OLS without Spatial Autocorrelation and Weighting

```
proc glm;
model z=pexposure pctage65p pctownhome;
title1 " ";
title2 "ordinary least squares";
run;
```

The GLM Procedure
Dependent Variable: Z

Source	DF	Sum of Squares	Mean Square	F Value	Pr > F
Model	3	28.6372808	9.5457603	22.10	<.0001
Error	277	119.6190948	0.4318379		
Corrected Total	280	148.2563756			

R-Square	Coeff Var	Root MSE	Z Mean
0.193161	-304.5867	0.657144	-0.215749

Parameter	Estimate	Standard Error	t Value	Pr > |t|
Intercept	-0.517276345	0.15855721	-3.26	0.0012
PEXPOSURE	0.048836274	0.03506349	1.39	0.1648
PCTAGE65P	3.950889564	0.60549832	6.53	<.0001
PCTOWNHOME	-0.560041340	0.17030802	-3.29	0.0011

The OLS is significant with R-squared 0.193 and p-value <.0001. PCTAGE65P and PCTOWNHOME are significant. Exposure does not. The residual standard error is $= \sqrt{(0.657)}$ on 277 degrees of freedom. There is information in the residuals that we need to extract. As such, we fit a model that takes this into account.

19.4.1.2 Model 2: Weighted Linear Regression without Spatial Autocorrelation

```
proc glm;
model z=pexposure pctage65p pctownhome;
weight pop8;
output out=stat2 p=fitw r=residw stdr=stdresidw student=studresidw;
title1 " ";
title2 "weighted least squares";
run;
```

The GLM Procedure

Dependent Variable: Z

Weight: POP8

Source	DF	Sum of Squares	Mean Square	F Value	Pr > F
Model	3	76580.9917	25526.9972	22.75	<.0001
Error	277	310778.3324	1121.9434		
Corrected Total	280	387359.3240			

R-Square	Coeff Var	Root MSE	Z Mean
0.197700	-8967.447	33.49542	-0.373522

Parameter	Estimate	Standard Error	t Value	Pr > |t|
Intercept	-0.778374202	0.14116037	-5.51	<.0001
PEXPOSURE	0.076256191	0.02731218	2.79	0.0056
PCTAGE65P	3.856560298	0.57126455	6.75	<.0001
PCTOWNHOME	-0.398694776	0.15305138	-2.60	0.0097

The weighted linear regression model has resulted in the TCE (PEXPOSURE) has become significant, indicating that tracts closer to the TCE sites have higher incidence. Thus, the information has moved from the residuals to the model.

```
proc print noobs;
var studresidw;
run;
```

We included the data set of the studentized residuals. Here is a subset:

studresidw
0.43616
0.29782
-0.76583
0.00944
0.84629
-0.18553
-0.60039
-0.21942

19.4.1.3 Model 3: OLS with Spatially Autocorrelated Errors

```
proc mixed data=one maxiter=250 ;
 model z = pexposure pctage65p pctownhome / solution outp=pred ;
 repeated / type=sp(sph)(x y) subject=intercept local ;
 parms (1) (5 to 50 by 5) (.5 to 1.5 by 0.25) /;
title1 "Results from Spatial OLS";
 title2 "unweighted";
 run;
```

As seen in the SAS code, the spatial autocorrelation is modeled by the Spherical (specified by "SPH") model with the spatial locations "X" and "Y". The parameter associated with this spherical model is specified by "Parms".

The Mixed Procedure

Model Information	
Data Set	WORK.ONE
Dependent Variable	Z
Covariance Structure	Spatial Spherical
Subject Effect	Intercept
Estimation Method	REML
Residual Variance Method	Profile
Fixed Effects SE Method	Model-Based
Degrees of Freedom Method	Between-Within

Dimensions	
Covariance Parameters	3
Columns in X	4
Columns in Z	0
Subjects	1
Max Obs per Subject	281

```
Convergence criteria met.
```

Covariance Parameter Estimates

Cov Parm	**Subject**	**Estimate**
Variance	Intercept	0.05582
SP(SPH)	Intercept	6.9267
Residual		0.3740

Incorporating the spatial autocorrelation, the estimated local variance (i.e. "Residual") is 0.3740, which is smaller than the 0.4318 in Model 1. The estimated range of spatial autocorrelation (i.e. "SP(SPH)") is 6.9267, which means that the observations within distance of 6.9267 are correlated. The added "sill" variance (i.e. "Variance") is 0.0558.

Fit Statistics	
-2 Res Log Likelihood	560.9
AIC (Smaller is Better)	566.9
AICC (Smaller is Better)	567.0
BIC (Smaller is Better)	577.8

PARMS Model Likelihood Ratio Test		
DF	Chi-Square	Pr > ChiSq
2	9.34	0.0094

Solution for Fixed Effects					
Effect	Estimate	Standard Error	DF	t Value	Pr > |t|
Intercept	-0.7222	0.1972	0	-3.66	.
PEXPOSURE	0.08266	0.04336	277	1.91	0.0576
PCTAGE65P	3.7093	0.6188	277	5.99	<.0001
PCTOWNHOME	-0.3245	0.2044	277	-1.59	0.1136

With adjusted spatial autocorrelation, exposure and percentage town home are not significant. The percentage over 65 iss still significant.

19.4.1.4 Model 4: Weighted LS with Spatially Autocorrelated Errors

```
proc mixed data=one maxiter=250 ;
  model z = pexposure pctage65p pctownhome / solution outp=pred ;
  repeated / type=sp(sph)(x y) subject=intercept local ;
  parms (1) (5 to 50 by 5) (.5 to 1.5 by 0.25) /;
  weight pop8;
  title1 "Model 4: WLS with spatial autocorrelation";
  title2 "weighted";
  run;
quit;
```

Model 4 extends Model 2 with weighting (i.e. "weight") with spatial autocorrelation.

Model Information	
Data Set	WORK.ONE
Dependent Variable	Z
Weight Variable	POP8
Covariance Structure	Spatial Spherical
Subject Effect	Intercept
Estimation Method	REML
Residual Variance Method	Profile
Fixed Effects SE Method	Model-Based
Degrees of Freedom Method	Between-Within

Dimensions	
Covariance Parameters	3
Columns in X	4
Columns in Z	0
Subjects	1
Max Obs per Subject	281

```
Convergence criteria met.
```

Covariance Parameter Estimates

Cov Parm	Subject	Estimate
Variance	Intercept	127.12
SP(SPH)	Intercept	6.8565
Residual		997.65

PARMS Model Likelihood Ratio Test

DF	Chi-Square	Pr > ChiSq
2	7.31	0.0258

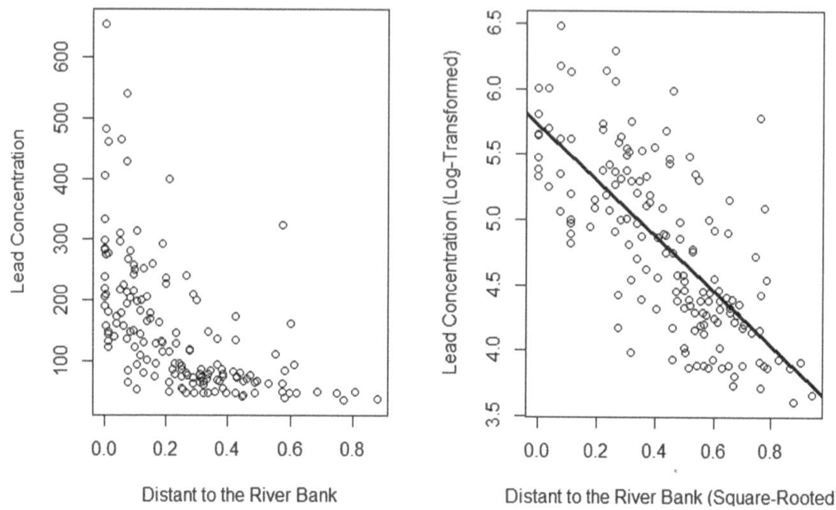

FIGURE 19.2
Scatter Plots between Lead Concentration and Distance from Meuse River Bank and Their Transformed Data to Identify the Linear Relationship. The Line is the Classical Regression Fitted Line.

Solution for Fixed Effects					
Effect	**Estimate**	**Standard Error**	**DF**	**t Value**	**Pr > |t|**
Intercept	-0.9161	0.1648	0	-5.56	.
PEXPOSURE	0.09563	0.03215	277	2.97	0.0032
PCTAGE65P	3.5763	0.5920	277	6.04	<.0001
PCTOWNHOME	-0.2285	0.1761	277	-1.30	0.1956

In this model 4, exposure and percentage over 65 are significant. The percent town home is not.

19.4.2 Spatial Analysis of Meuse River Data Using R

19.4.2.1 Graphical Analysis

With respect to the Meuse River data, there appears to be a spatial trend, as seen in Figure 19.1. A plot of the data presents a reasonable relationship for modeling. In Figure 19.2 (left panel), there is a clear nonlinear relationship between the lead concentration and the distance to the Meuse riverbank. A log-square rooted transformation (Bivand et al. 2008) produces a reasonable linear relationship to model the lead concentration to the distance, as shown in the right panel of Figure 19.2.

With this graphical analysis, let us use the log transformation of the lead concentration and denote it as "tlead". Similarly, we take the square root transformation of the distance and denote it as "tdist".

19.4.2.2 Linear Regression without Spatial Autocorrelation

Using the transformed data, consider the classical regression model:

$$\text{tlead}(s) = \beta_0 + \beta_1 \times \text{tdist}(s) + \varepsilon(s) \tag{19.1}$$

where s represents the 155 observation locations as seen in Figure 19.1. The model is fitted by several statistical software programs including the R function "lm" in the R code chunk:

```
# Transformed the data
>meuse$tlead = log(meuse$lead)
>meuse$tdist = sqrt(meuse$dist)
# Run a linear regression
>lead.lm = lm(tlead~tdist, meuse)
# print out the summary for the linear regression
>summary(lead.lm)
```

The R code produces the following summary of linear regression fit:

```
Coefficients:        Estimate Std. Error  t value Pr(>|t|)
(Intercept)     5.72902   0.08094  70.78  <2e-16 ***
tdist         -2.12004   0.16521  -12.83  <2e-16 ***
Residual standard error: 0.464 on 153 degrees of freedom
Multiple R-squared:  0.5184,    Adjusted R-squared:  0.5152
F-statistic: 164.7 on 1 and 153 DF,  p-value: < 2.2e-16
```

The linear regression model is statistically significant with F-Statistic = 164.7 on 1 and 153 degrees of freedom and p-value < 0.0001. The estimated regression equation is and the estimated residual standard error $\hat{\sigma}$=0.464 (or $\hat{\sigma}^2$ =0 0.215).

As customary in the typical regression diagnostics, we check the assumptions of homogeneity and normally distributed through the residuals.

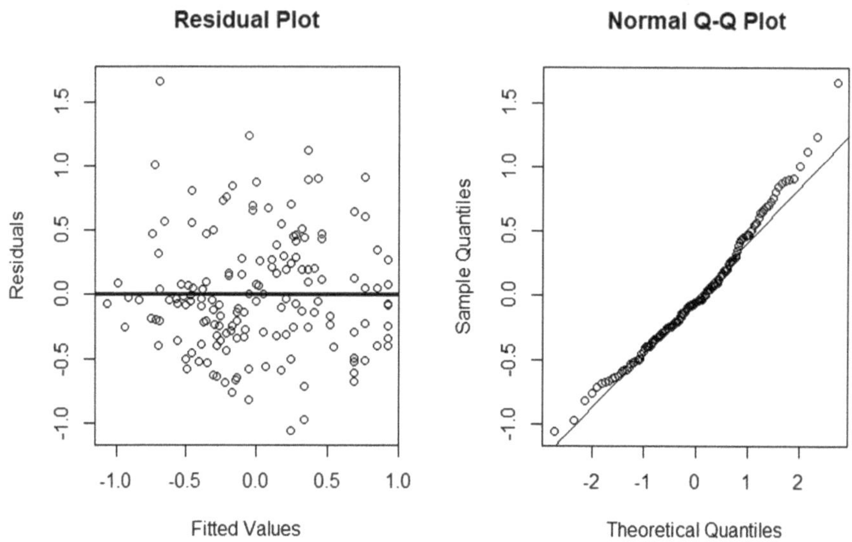

FIGURE 19.3
Regression Model Diagnostics.

These residuals are graphically displayed in Figure 19.3. The residual plot and the normal QQ plot are obtained using R code chunk:

```
># Extract the fitted values and center them
>meuse$sfitted = predict(lead.lm,meuse)-mean(predict(lead.lm, meuse))
># Extract the residuals
> meuse$residuals = residuals(lead.lm)
# Residual plot for model diagnositics
>par(mfrow=c(1,2))
>plot(residuals~sfitted, meuse, xlab="Fitted Values",
     ylab="Residuals", main="Residual Plot")
>abline(h=0, lwd=3)
>#QQ-normal plot for residuals
>qqnorm(meuse$residuals)
>qqline(meuse$residuals)
```

We check the independence assumption through the residual plot. We plot the residuals at the spatial locations, as shown in Figure 19.4, using the "ssplot" function in R:

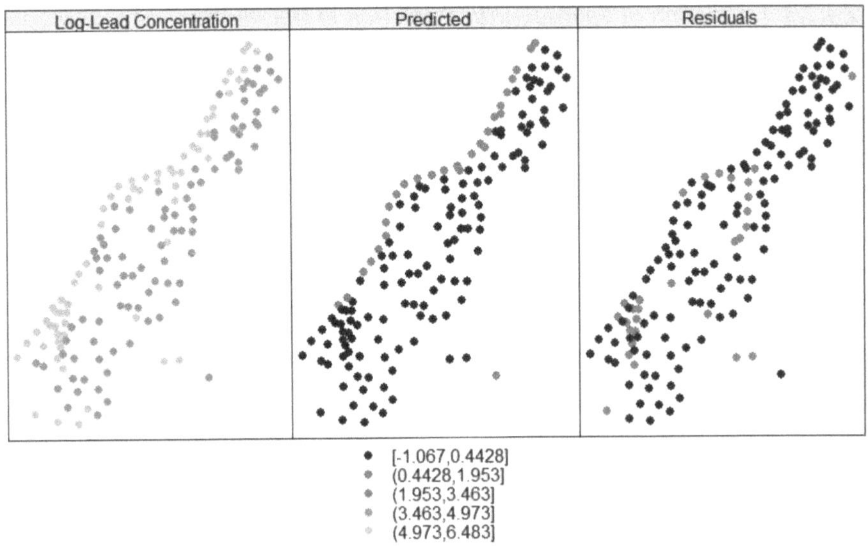

FIGURE 19.4
Data with Predicted and the Associated Residuals.

```
># spatial plot
>spplot(meuse, c("tlead","sfitted","residuals"),
names.attr=c("Log-Lead Concentration", "Predicted","Residuals"))
```

The left panel of Figure 19.4. shows the spatial distribution of lead concentration, the middle panel shows the linear regression predicted lead (log-) concentration, and the right panel shows the residuals from the regression model.

In Figure 19.4, the middle panel that the spatial trend from the left panel is reasonably modeled and largely removed. However, the residuals displayed in the right panel seem spatially clustered with similar values close to each other (i.e. the residuals are spatially correlated and they are not independent), which violates the regression assumption of independent data.

19.4.2.3 Estimation of Spatial Autocorrelation in the Residuals

To decide if the spatial autocorrelation existed in the residuals, we plot all the pairs from the residuals in modeling lead concentrations to their spatial separation distance. This is implemented using R function "hscat" from the R library "gstat" as

```
># load the "gstat" library
>library(gstat)
># display spatial correlation by lag in distance
>hscat(tlead~tdist, meuse, (0:8)*100)
```

Figure 19.5 shows that the spatial correlations are 0.722, 0.587, 0.431, 0.286,0.194, 0.105,0.006, -0.0885 at the distance classes of 0 to 100, 100 to 200, 200 to 300, 300 to 400, 400 to 500, 500 to 600, 600 to 700, 700 to 800. This confirms a typical pattern in spatial autocorrelation, where the closer the spatial distance, the higher the spatial autocorrelation.

An alternative approach to estimate the spatial autocorrelation is to use the variogram [The variogram characterizes the spatial continuity or

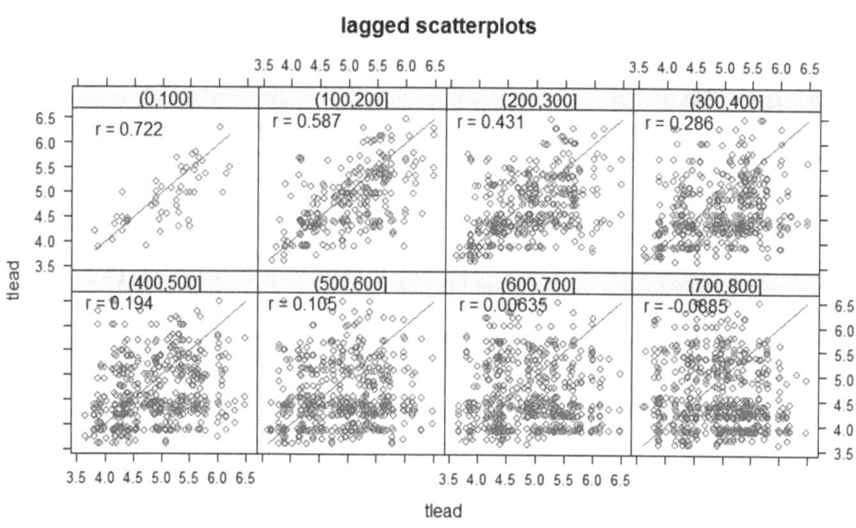

FIGURE 19.5
Spatially Lagged Scatterplot to Show Spatial Autocorrelation.

roughness of a data set.] to model the variance function as a function of the spatial distance. Intuitively in variogram modeling, the variance (i.e., called the "semivariogram" in spatial statistics) function increases as spatial distance increases, in contrast to spatial autocorrelation, where the spatial autocorrelation decreases as spatial distance increases. This function calculates all the semivariogram pairs in the residuals and plots all the pairs to their spatial distance, as seen in Figure 19.6. In that figure, a large spread of these semivariogram pairs exists]. This is done by examining the variogram and variogram cloud. The variogram cloud is produced by plotting all possible paired squared differences to their spatial distances, which is computationally extensive, but is implemented in R with R function "variogram" as

```
> plot(variogram(tlead~tdist,meuse, cloud=T))
```

To smooth the variations in the variogram cloud, as shown in Figure 19.6, one takes the averages of semivariogram cloud values on spatial distance intervals to provide a more smoothed representation of the semivariances. This is done using the same "variogram" function in R as

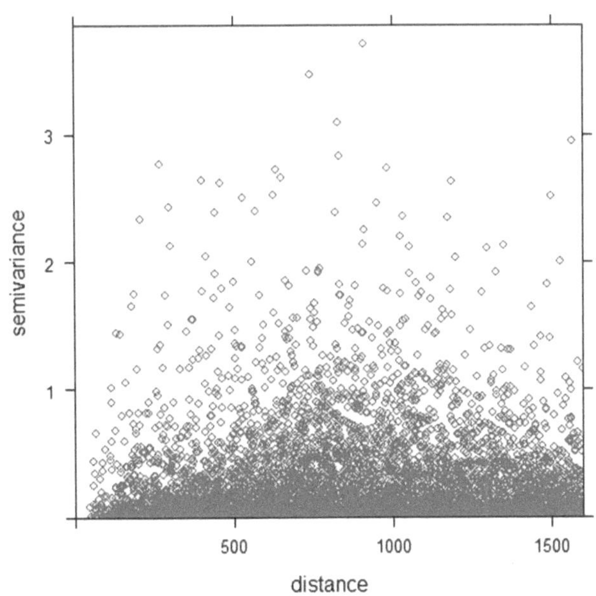

FIGURE 19.6
Variogram Cloud from the Residuals.

```
># the empirical variogram as average the variogram cloud
>var.lead  = variogram(tlead~tdist, meuse)
>#plot the variogram
>plot(var.lead)
```

This produces the empirical variogram (i.e., dots) in Figure 19.7. In the empirical variogram, the obvious increasing relationship for the semivariance along with their spatial distance confirms that the residuals from the linear regression in Equation (19.1) are not independent and spatial autocorrelation exists in the residuals.

Due to the large variation in the variogram cloud and the associated empirical variogram, a family of theoretical variogram models is proposed with the property that the variance function should be a monotonically increasing function (or decreasing in spatial autocorrelation) with the increasing spatial distance. The theoretical variogram model is then used to fit the empirical variogram to be used for spatial statistical modeling. Among all the proposed theoretical variogram models, the most commonly used is the Materm variogram family, the powered exponential variogram family, and the spherical variogram family, as discussed in Cressie (1991, 2015). The line in Figure 19.7 is the fitted spherical variogram model (i.e., a three-parameter variogram model), which provides a reasonable fitting to the empirical variogram. This fitting is done by calling R function "fit.variogram" as follows:

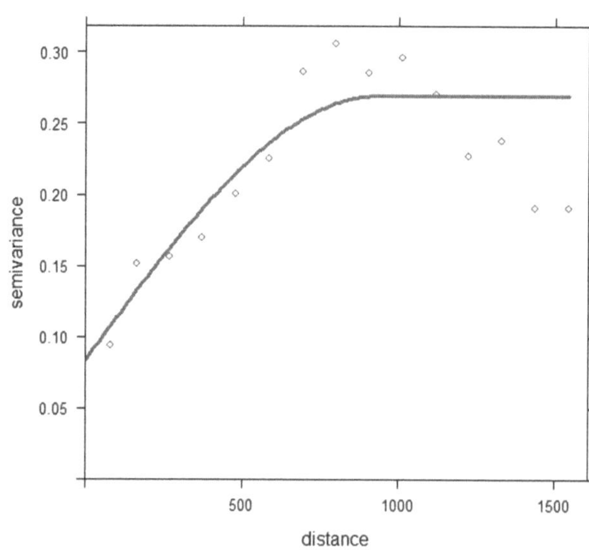

FIGURE 19.7
Plot for Empirical Variogram (Dots) and the Fitted Spherical Variogram Model (Line).

```
># Variogram fitting using spherical model "Sph"
>fit.lead = fit.variogram(var.lead, vgm("Sph"))
># add the fitted spherical variogram to the empirical variogram
>plot(var.lead, fit.lead, lwd=3)
># print the spherical variogram fiting
>fit.lead
model    psill   range
1  Nug 0.08345555  0.0000
2  Sph 0.18587945 927.7763
```

We fit the theoretical spherical variogram as specified as vgm("Sph") and estimate the nugget effect parameter (i.e., a discontinuity at the origin in the variogram. It is called the local variance or measurement error variance) at 0.083 and sill effect parameter (i.e., the variance of the observations) at 0.186. The range parameter is estimated at 927.776, which is the maximum spatial distance where the variance function will reach the asymptote, indicating no spatial autocorrelation beyond this range. We encourage the readers to fit other theoretical variogram models to these data.

19.4.3 Geostatistical Regression Model

Due to the spatial dependence in the residuals, as shown in Section 19.4.2.3, we would need to refit the regression model in Section 19.4.2.2 to incorporate the spatial autocorrelation. In this case, we re-estimate the regression parameters (such as β_0 and β_1 in Equation 19.1) along with the covariance parameters from the theoretical variogram model (such as the nugget, sill, and range parameters in the spherical variogram model).

The commonly used least-squares estimation in the linear regression is not appropriate anymore, and the maximum likelihood estimation (MLE) should be used. With MLE, we can construct the likelihood function to include the regression parameters from the regression component along with the parameters from the variance function in the residuals. This likelihood function is then to be maximized to get the parameter estimates and other associated statistical inferences (see Cressie 1991, 2015; Diggle and Ribeiro 2007). We make use the R function "likfit" from R package "geoR" for this illustration with following R code chunk:

```
># make geodata to be used for "likfit"
>meuse.geodata = as.geodata(meuse, data.col = "tlead",
     covar.col="tdist")
># call "likefit" to fit the geostatistical regression
> lead.spatial = likfit(meuse.geodata, trend=~tdist,
     ini = c(1,800), cov.model = "sph", fix.nugget=F)
> # print the MLE fitting
> lead.spatial
likfit: estimated model parameters:
     beta0        beta1       tausq      sigmasq          phi
"   5.6568" " -2.0045" "  0.1008" "  0.1203" "796.0994"
      likfit: maximised log-likelihood = -80.42
```

We can extract the estimates of regression parameters and the associated variances to construct z-statistics and the associated p-values for statistical inference as follows:

```
># extract the parameter estimates
>beta.est = lead.spatial$beta
># extract the estimated variance
>beta.se = diag(sqrt(lead.spatial$beta.var))
>beta.se
># calculate the z-statistics
>zval = beta.est/beta.se;zval
```

These values are listed in the right column, "Geostatistical Regression," of Table 19.2. In this table, the linear relationship between the log-transformed lead concentration and the square root of the distance to the Meuse riverbank is still statistically significant.

For the convenience of comparison, we summarize the output from the "Classical Regression" in Section 19.3.1 and "Residual Variogram Model" in Section 19.3.2. The estimated regression parameters are very close even though they are not identical. The estimated standard errors for the "Classical Regression" are smaller than those

TABLE 19.2

Comparison of the parameter estimation from the model fitting

Parameter Estimate	Classical Regression	Residual Variogram Model	Geostatistical Regression
β_0	5.729(0.081)***	NA	5.657(0.139)***
β_1	−2.120(0.165)***	NA	−2.005(0.262)***
σ^2 (Sill)	0.215	0.186	0.120
τ^2 (Nugget)	NA	0.083	0.101
□ (Range)	NA	927.776	796.099

for the "Geostatistical Regression," which is intuitively true since the "Classical Regression" ignores the dependence from the data, and as a consequence, erroneously deflates the variance estimates. Therefore, the "Classical Regression" inflates the type-I error rate to conclude statistical significance. In addition, the estimated residual variance (i.e., 0.120) is smaller in the "Geostatistical Regression" than in the "Classical Regression" (i.e., 0.215), where part of the variance is explained by the spatial structure in the variogram.

19.5 Discussions and Conclusions

In this chapter, we provided an introductory overview of spatial statistics from the perspective of regression modeling. We presented a remedy for using spatial data with classical regression modeling, a common approach in public health application. We used the leukemia data with heavy metal concentrations as the continuous response measures, thus classical regression. While we focused on continuous response, other types of response can extend the generalized linear model to incorporate the spatial dependence using the MLE. The implementation in R is done in the package "geoRglm" (https://cran.r-project.org/web/packages/geoRglm/index.html).

There are several topics in spatial statistics we did not cover in this chapter (for example, spatial prediction). When modeled with the variogram model, spatial data is used to predict the values at unobserved locations. This spatial prediction, referred to as kriging, is available in all software. Another example is disease mapping. The mapping of disease incidence and prevalence of diseases is particularly important in public health research, as extensively discussed in Lawson (2006, 2013, 2016). Several R packages are available at Roger Bivand's CRAN Task View (https://cran.r-project.org/web/views/Spatial.html).

For software development and implementation, SAS has developed several procedures for spatial analysis, which are available online at https://support.sas.com/rnd/app/stat/procedures/SpatialAnalysis.html:

1. "KRIGE2D" to perform ordinary kriging or spatial prediction for spatial point referenced data.
2. "SIM2D" to produce a spatial simulation for a Gaussian random field with a specified mean and covariance structure in two dimensions by using an LU decomposition technique.
3. "SPP" to analyze spatial point patterns.
4. "VARIOGRAM" to compute variogram diagnostics to describe the spatial covariance structure in spatial point referenced data.

In R, there are several R packages developed for spatial statistics. The R package "sp" developed by Bivand et al. (2008) can be downloaded from https://cran.r-project.org/web/packages/sp/index.html with online documentations. Another R package from this group is the "gstat" for "Spatial and Spatio-Temporal Geostatistical Modelling, Prediction and Simulation," which is detailed in www.gstat.org.

The R package "geoR" (https://cran.r-project.org/web/packages/geoR/index.html), developed by Paulo J. Ribeiro Jr and Peter J. Diggle, is for the analysis of geostatistical data including the traditional, likelihood-based, and Bayesian methods. The R package "geoRglm" (https://cran.r-project.org/web/packages/geoRglm/index.html), developed by Ole F. Christensen and Paulo J. Ribeiro Jr., is an extension of the package "geoR" for generalized linear spatial models. The Bayesian posterior and predictive inference is based on Markov chain Monte Carlo methods. The CRAN Task View maintained by Roger Bivand (https://cran.r-project.org/web/views/Spatial.html) also lists all the developments in R for spatial statistics.

The results presented above showed that spatial correlations exist among schools in Louisiana with respect to math achievement, meaning that those in close proximity to one another are more similar than would be expected by chance. They may serve students from the same families and neighborhoods, and may experience similar processes with respect to recruiting and retaining teachers and administrators. These relationships persist even when student and school contextual variables are incorporated in the model examined. They suggest that a potentially valuable avenue of future research related to poverty and educational is to explicitly incorporate location in models of schooling processes. For example, the interstate roadway system in Louisiana was not imposed in Figure 19.3, yet it appears that outside of urban centers, the distribution of failing schools is related to the presence of this interstate roadway. These other hypotheses can be explored with spatial modeling techniques.

19.6 Exercises

Use the Leukemia data and consider additional covariates in "Tractcas: number of cases 1978–1982" and "Avgdist: average distance between centroid and TCE sites" and conduct a spatial analysis.

References

Anselin, L. (1995). Local indicators of spatial association – LISA. *Geographical Analysis*, 27(2), 93–115.
Bivand, R. S., Pebesma, E. J. and Gomez-Rubio, V. (2008). *Applied Spatial Data Analysis with R*. New York: Springer.
Cressie, N. A. C. (1991). *Statistics for Spatial Data*. New York: Wiley.
Cressie, N. A. C. (2015). *Statistics for Spatial Data* (Revised 2nd Edition). New York: Wiley.
Diggle, P. J. and Ribeiro, P. J. (2007). *Model-Based Geostatistics*. New York: Springer.
Fotheringham, A., Brunsdon, C., and Charlton, M. (2000). *Quantitative Geography: Perspectives on Spatial Data Analysis*. London: Sage.
Haining, R. (1990). *Spatial Data Analysis in the Social and Environmental Sciences*. Cambridge University Press.
Lawson, A. B. (2006). *Statistical Methods in Spatial Epidemiology*, 2nd ed. New York: Wiley.
Lawson, A. B. (2013). *Bayesian Disease Mapping*. New York: CRC Press.
Lawson, A. B. (2016). *Handbook of Spatial Epidemiology*. New York: CRC Press.
Sudipto, B., Bradley, C. P. and Gelfand, A. E. (2014). *Hierarchical Modeling and Analysis for Spatial Data*, 2nd ed. Boca Raton, FL: Chapman and Hall/CRC.
Xu, Y. and Kennedy, E. (2015). An introduction to spatial analysis in social science research. *The Quantitative Methods for Psychology*, 11(1), 22–31.

20

Structural Equation Modeling

20.1 Research Interest/Questions

Structural equation modeling (SEM) is a powerful but general statistical technique combining factor analysis and regression or path analysis. The theoretical constructs of SEM are presented by latent factors, while the relationships between the theoretical constructs are shown by path or regression coefficients. SEM is often visualized by a graphical path diagram. The statistical model is presented in a set of equations. In particular, SEM encompasses a broad array of models, from linear regression to measurement models to simultaneous equations. It is not just an estimation method for a particular model, but it involves a way of thinking, a way of writing, and a way of estimating.

An extensive literature on SEM exists, including works by Byrne (2012), Bollen (1989), Kline (2011), and Westland (2015), to list a few. For applications, there are various software implementations of SEM exist, including Mplus, SAS, R, and Stata with online trainings, documentation, videos, and presentations. A recent book on SEM with R and SAS can be found by Chen and Yung (2023), which promoted a step-by-step approach with real data analysis using R package *lavaan* and SAS procedure *proc calis*. Related theoretical background in SEM can also be found in this book.

In this chapter, we provide introductory applications to SEM using Mplus, SAS R, and Stata, with the example from Shook-Sa et al. (2017). This example assessed the links between tobacco smoke exposure, volatile organic compounds (VOCs), and asthma respiratory function (RF) in adolescents.

20.2 Asthma Data

Asthma is an inflammatory airway disease that affects 22 million Americans and is the most common chronic disease in young children. The incidence of

DOI: 10.1201/9781003596370-9

asthma has been growing over the past several decades (Stapleton et al. 2011) and is a growing public health concern because of its prevalence, significant health risks, and high healthcare costs (Nurmatov et al. 2013). Researchers have found associations between impaired RF, including asthma and increased symptoms among asthmatics, and common indoor air pollutants, including tobacco smoke exposure and VOCs. However, the findings linking VOC exposure and asthma are inconsistent. The studies have different designs and limitations, including challenges measuring VOC exposure, small sample sizes, and suboptimal statistical methodologies. The correlation between tobacco smoke exposure and VOCs, and associations between both tobacco smoke and VOCs with RF, led Shook-Sa et al. (2017) to test a SEM to model the causal relationship among these risk variables, hypothesized in Figure 20.1. The model predicts RF based on tobacco smoke exposure and VOCs (xylene/ethylbenzene and toluene), controlling for the effects of age, height, weight, and poverty status on RF and poverty status on tobacco smoke exposure and VOCs. Figure 20.1 is a path diagram (A path diagram is made up of rectangles and circles). The arrows from the factors to the variables represent linear regression coefficients of factor loadings (interpreted as estimates of the true population loading values of those indicators for that particular latent variable). Observed variables are represented by rectangles, and latent or unmeasured factors are depicted by a circle or ellipse, which are not actually measured, and serve as a construct in the model. Single-headed arrows represent a causal relationship, with the variable at the head of the arrow as the causing variable. Double-headed arrows indicate covariance or correlations without a causal interpretation. In Figure 20.1, the path diagram has oval shapes representing latent variables. We analyze the latent constructs, using some measured variables (or indicators).

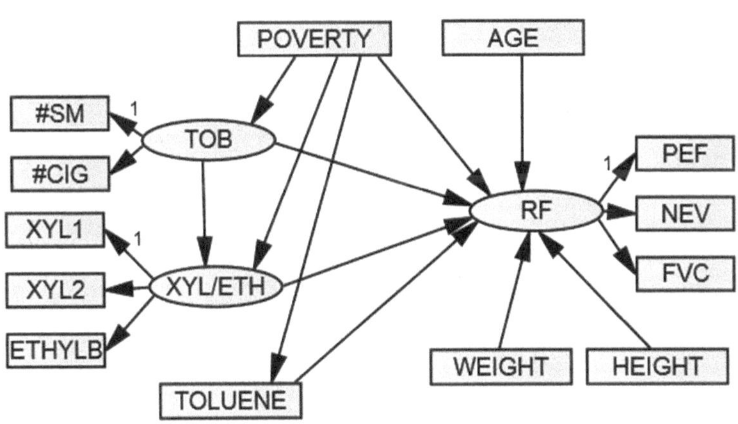

FIGURE 20.1
Respiratory Function, Tobacco Smoke Exposure, and VOCs.

We examine data from the 2011–2012 National Health and Nutritional Examination Surveys (NHANES), which is a nationally representative in-person survey of the noninstitutionalized civilian resident population of the United States that focuses on health and nutrition. The NHANES includes a number of items regarding RF, asthma, and other respiratory diseases, including physical assessments and self-reported data from respondents. A subset of the data made available for this analysis includes three spirometry measures of lung function: peak expiratory flow (PEF), extrapolated volume (NEV), and forced vital capacity (FVC), which is tobacco use by the respondent and tobacco exposure in the respondent's home. These data are presented in Table 20.1. In the 2011–2012 NHANES, metabolites for the following VOCs are examined in the urine samples: xylenes are 2-methylhippuric acid, 3-methipurc acid, and 4-methipurc acid; ethylbenzene is phenylglyoxylic acid; and toluene is N-acetyl-S-(benzyl)-L-cysteine. The covariates are age, poverty status, height, and weight. Shook-Sa et al. (2017) conducted an analysis for a sample of 748 of these observations. Due to missing values and for illustration purposes in this chapter, we use the 616 complete observations. Therefore, the reported parameter estimates and the values of model selection criteria listed in this chapter differ slightly from the original results in Shook-Sa et al. (2017).

20.3 Structural Equation Modeling

The associated SEM parameters are estimated statistically in a simultaneous modeling framework with the observed data. If the model meets the goodness-of-fit criteria, the hypothesized SEM provides adequate guidance to the structural causal effects among the variables. It is used as a guide in public health decision-making.

SEM is a system of statistical models to take a confirmatory approach to analyze the structural theories of public health, sociology, psychology, and other relevant scientific fields. It is typical that these "structural theories" represent theory-guided causal effects that generate data from multiple variables. Thus, SEM is used to model the causal processes from a series of structural equations, typically represented through regression-based models on the hypothesized theory. These equations are graphically depicted for a clearer conceptualization of the theory under study. The hypothesized theoretical causal model is then tested for its validity.

TABLE 20.1

Illustration of Ten Observations

RIDAGEYR	POVERTY	WTSA2YR	HEIGHT	WEIGHT	NSMOKE	NCIGH	NEVS	PEFS	FVCS	ETHLBENS	XYLENE1S	XYLENE2S	TOLUENE
15	5.00	43867.45	16.87	5.56	0	0	8.69	12.89	11.27	8.09	9.37	9.51	9.42
6	0.30	56168.36	12.47	2.37	0	0	10.23	7.95	8.02	7.78	9.40	9.51	9.42
18	1.78	46980.38	18.69	9.60	0	0	10.50	16.43	14.89	7.89	9.44	9.55	9.43
16	0.54	32816.14	17.29	5.37	0	0	10.83	12.83	12.24	7.78	9.37	9.50	9.43
10	5.00	24152.28	13.77	2.92	0	0	7.81	7.28	8.64	8.08	9.38	9.53	9.43
8	1.02	40782.51	13.07	3.12	0	0	9.02	8.98	9.56	7.92	9.39	9.50	9.43
16	1.64	35339.58	16.31	5.77	0	0	9.83	11.44	11.77	8.19	9.36	9.49	9.43
6	4.52	48992.62	12.00	2.29	0	0	7.34	6.86	7.16	8.41	9.61	9.60	9.43
8	1.07	46991.95	13.30	2.69	0	0	7.61	8.79	8.32	8.22	9.46	9.54	9.43
16	5.00	251258.89	16.49	6.12	0	0	10.23	11.68	11.65	8.33	10.25	9.73	9.43

20.4 Confirmatory Factor Analysis: Validity of Latent Constructs

The essence of SEM is to model the theoretical latent constructs that are not directly observed (i.e., called a "latent" construct). Therefore, these latent constructs would have to be validated before full SEM. In this sense, the SEM model extends the classical statistical regression modeling from previous chapters, which only uses the observed independent variables to model the observed dependent variables, to include observed and latent variables as both independent and dependent variables. For example, in Figure 20.1, aside from the observed variables (i.e., age, weight, height, poverty, Toluene), the RF is an unobserved latent construct that is constructed from three observed spirometry measures of lung function from PEF, NEV, and FVC. Also, the latent construct of "Tobacco" effect is constructed by the observed average

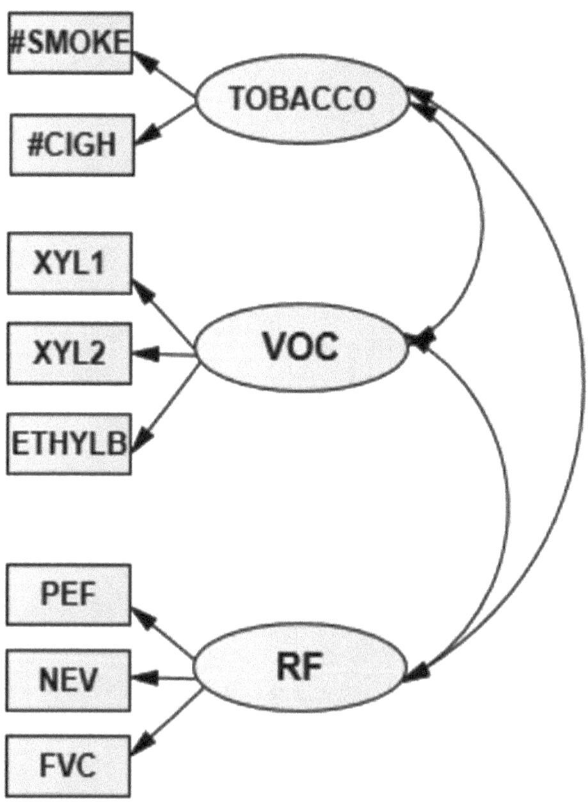

FIGURE 20.2
CFA Model Three Latent Constructs to Eight Observed Variables.

number of cigarettes smoked in the home per day (#CIGH) and the observed number of smokers in the household (#SMOKE). Similarly, the latent construct of VOCs is constructed by the observed 2-methylhippuric acid (XYL1), 3-methipurc acid and 4-methipurc acid (XYL2), and phenylglyoxylic acid (ETHYLB). This is graphically depicted in Figure 20.2.

Due to these latent structures, we need to test the construct validity from observed variables to the latent construct using confirmatory factor analysis (CFA) to be used for SEM. This is used to examine the covariance among the observed set of measures to gather information on their underlying latent construct. We test this hypothesis structure statistically. The CFA model is represented in Figure 20.2. Factor analysis assumes that the covariance between a set of observed variables is explained by a smaller set of underlying latent factors. In exploratory factor analysis, there is no prior hypothesis about the number of latent factors or the relationship between observed variables and the latent factor. Statistical procedures tell the number of factors and provide methods to estimate the factor loadings.

In structural modeling, the CFA is used to fit the data to test the validity of latent constructs before the full SEM:

1. To obtain estimates of the factor loadings, the variances and covariances of the factor, and the residual error variances of the observed variables.
2. To assess the fit of the model.

20.5 SEM Versus Regression Analysis

A typical regression analysis is at the predictor–outcome framework, commonly known as "X predicts Y". Mathematically and statistically, this idea is represented in the simple linear regression analysis, as

$$Y = \beta X + \epsilon$$

$$\text{Outcome} = \text{Parameter} * \text{Predictor} + \epsilon$$

$$Y \leftarrow ----X$$

where β is called the effect, regression coefficient, or path coefficient. An error term ϵ is added to show that the prediction of Y from X is not perfect. Graphically, we present $Y \leftarrow X$ as the path in the diagram.

There are differences between SEM and regression analysis. SEM is more complicated than multiple regression analysis for multiple predictor–outcome relationships. We summarize these differences as

1. More variables (not just X and Y)
2. More equations or functional relationships (not just $X \rightarrow Y$)
3. Correlated errors, system of equations can have correlated errors
4. Direct and indirect effects
5. Latent variables
6. Parametric constraints
7. Multiple-group analysis

SEM is known by many other names, but fundamentally, they are the same. Some of these names are path analysis (attributed to Sewall Wright), LISREL model (JKW model), covariance structures analysis, analysis of moment structures, CFA, and causal modeling, among others.

20.6 Analysis of NHANES Data

We demonstrate the fitting with three factors and eight observed variables, as shown in Figure 20.2. The RF is an unobserved latent construct, which is constructed from three observed spirometry measures of lung function from PEF, NEV, and FVC. The latent construct of "Tobacco" effect is constructed by the observed average number of cigarettes smoked in the home per day (#CIGH) and the observed number of smokers in the household (#SMOKE). Similarly, the latent construct of VOCs is constructed by the observed 2-methylhippuric acid (XYL1), 3-methipurc acid and 4-methipurc acid (XYL2), and phenylglyoxylic acid (ETHYLB). We demonstrate the fit with MPlus, SAS, and R.

20.6.1 Analysis of NHANES Data Using MPus Program

A standalone commercial software for SEM is "Mplus," which can be accessed from www.statmodel.com/. The Mplus demo version can be downloaded freely from this website and includes all of the capabilities of the regular version of Mplus. The only limitation is the number of observed variables that can be used in an analysis. From the homepage, one may access training materials, handouts, and videos from training courses and international conferences. In addition, the online Mplus user's guide contains several thousands of pages, which include the detailed descriptions of SEM models, Mplus implementation with Mplus code, and data for users to practice. The textbook by Byrne (2012) is an excellent reference on SEM.

The Mplus code fitting the model is as follows with the first on CFA analysis and the second on the full SEM.

20.6.1.1 SEM Program for Chapter 20-CFA Analysis

The "Title" is an optional command to name the analysis. This command can be omitted from Mplus code, but it is the place to record what this analysis is about and to keep track of your sequential analysis. In the above code, we name it as "SEM Program for Chapter 20-CFA analysis" to denote this is for CFA analysis.

Data:

```
File is "nhanesyouthNew.dat";
```

The "DATA" command is a required command to link the dataset to be analyzed. In this example, the dataset is named "nhanesyouthNew.dat," which is stored at the same file fold with 616 observation and no missing values.

Variable:

```
Names are RIDAGEYR POVERTY WTSA2YR HEIGHT WEIGHT NSMOKE NCIGH
   NEVS PEFS FVCS ETHLBENS XYLENE1S XYLENE2S TOLUENE;
```

The "VARIABLE" is a command to identify the variables in the data file, which should be consistent with the "DATA" command, along with its NAMES option. We make use "USEVARIABLES ARE" to identify the variables to be used for this analysis, such as NCIGH, NSMOKE, XYLENE1S, XYLENE2S, ETHLBENS, PEFS, NEVS, FVCS.

```
USEVARIABLES ARE
NSMOKE NCIGH NEVS PEFS FVCS ETHLBENS XYLENE1S
XYLENE2S;
WEIGHT IS WTSA2YR;
```

The "Weight" is to include the survey weights from NHANES survey sampling into SEM model fitting.

```
ANALYSIS:
   ITERATIONS=50000;
   ESTIMATOR IS MLR;
```

The "Analysis" is to identify analysis options and methods. In this Mplus code, we run 50,000 iterations "ITERATIONS=50000" and use "ESTIMATOR IS MLR," where "MLR" is the maximum likelihood robust estimator to account for non-normally distributed data and the missing data.

```
MODEL:
   TOBACCO BY NCIGH NSMOKE;
   NSMOKE@0;
   VOC BY XYLENE1S XYLENE2S ETHLBENS;
```

The "Model" command is the most important command to specify the hypothesized SEM to be fitted. Three key words, "BY", "ON" and "WITH", are required to specify the SEM The first option "BY" is used to0063 define the regression link between the underlying latent construct measured by its related observed variables. Corresponding to the Mplus code above, this can be seen from the above Mplus code that the latent variable "TOBACCO" is measured "**BY** NCIGH NSMOKE", the latent variable "VOC" is measured "**BY** XYLENE1S XYLENE2S ETHLBENS" and the latent variable "RF" is measured "**BY** PEFS NEVS FVCS". The "NSMOKE@0" is to set the error variance to be zero due to the estimated negative error variance in the initial SEM fitting. The second option "ON" is an abbreviation of "regressed on" to define any regression paths in the full SEM model which will be seen in next section. The third option "WITH" is short for "correlated with" which is used to identify covariance (i.e., correlational) between latent variables in either the measurement or structural models. These options are not used here for CFA analysis, but will be illustrated in the full SEM model.

```
RF  BY PEFS NEVS FVCS;
OUTPUT:
MODINDICES;
```

The "OUTPUT" command is an option command to output information needed from the data and model fitting. In the above Mplus code, we output the "MODINDICES," which is the "MODEL MODIFICATION INDICES" to be used for model specifications.

When the above Mplus code is executed, the Mplus will output a long list of the model fitting information and the associated parameter estimates. For explanation purposes, we edit out some redundant information and only keep the key information below to be explained.

```
THE MODEL ESTIMATION TERMINATED NORMALLY
MODEL FIT INFORMATION
Chi-Square Test of Model Fit
      Value                                38.959*
        Degrees of Freedom                   18
        P-Value                             0.0029
RMSEA (Root Mean Square Error Of Approximation)
         Estimate                            0.043
         90 Percent C.I.                   (0.025  0.062)
         Probability RMSEA <= .05           0.693
CFI/TLI
         CFI                                0.946
         TLI                                0.916
SRMR (Standardized Root Mean Square Residual)
         Value                              0.036
```

The first goodness-of-fit statistic is the Chi-Square Test of Model Fit, calculated as the sample size minus 1, times the minimum fit function. The Chi-square statistic is 38.959 with 18 degrees of freedom and a p-value of 0.0029, which indicates that the fit of the data to the hypothesized CFA model is not entirely adequate. This chi-square statistic can be misleading. Another GOF statistic is the comparative fit index (CFI; Bentler 1990). The CFI is typically normalized from zero to

1.00, with values close to 1.00 for well-fitting models. A revised cutoff value close to .95 has more recently been advised (Hu and Bentler 1999). Another is the is the Tucker-Lewis fit index (TLI; Tucker and Lewis 1973). Others are the root mean square error of approximation (RMSEA; Steiger and Lind 1980) and the standardized root mean square residual (SRMR), where the RMSEA <0.05 indicates good fit, and values as high as .08 represent reasonable errors of approximation in the population (Browne and Cudeck 1993).

In SEM fitting, the objective is to find parameter values, so as to minimize a discrepancy function, from the residuals between the sample covariance matrix calculated from the observed data and the population covariance matrix implied by the model. When this is achieved, the minimal value, Fmin, is then used to construct the goodness-of-fit (GOF) statistics.

For the asthma data, the RMSEA is 0.043, CFI is 0.946, and TLI is 0.916, which indicate the CFA model in Figure 20.2 fits the data adequately, and therefore validates the three latent constructs from the eight observed variables.

20.6.1.2 Using the Fitted CFA Model for Full SEM

Following the establishment of well-fitted CFA measurement models of the three latent variables (tobacco exposure, xylene/ethylbenzene, and RF) from the eight observed indicators, an SEM in Figure 20.1 is then fitted to assess the relationships between RF, tobacco smoke exposure, and VOCs among adolescents, controlling for covariates (age, height, weight, and poverty status). The Mplus code is

```
Title:
  Full SEM for Chapter 20
Data:
  File is "nhanesyouthNew.dat";
Variable:
  Names are RIDAGEYR POVERTY WTSA2YR HEIGHT WEIGHT NSMOKE
            NCIGH NEVS PEFS FVCS ETHLBENS XYLENE1S XYLENE2S
            TOLUENE
    ;
```

```
USEVARIABLES ARE
   NCIGH NSMOKE XYLENE1S XYLENE2S ETHLBENS PEFS NEVS
           FVCS RIDAGEYR POVERTY TOLUENE HEIGHT WEIGHT;
           WEIGHT IS WTSA2YR;
      ANALYSIS:
           ITERATIONS=50000;
           ESTIMATOR IS MLR;
      MODEL:
           TOBACCO BY NCIGH NSMOKE;
           NSMOKE@0;
           VOC BY XYLENE1S XYLENE2S ETHLBENS;
           RF BY PEFS NEVS FVCS;
           RF ON TOLUENE VOC TOBACCO RIDAGEYR HEIGHT
                 WEIGHT POVERTY;
           VOC ON POVERTY;
           TOBACCO ON POVERTY;
           TOLUENE ON POVERTY;
           VOC ON TOBACCO;
      OUTPUT:
           MODINDICES;
```

The Mplus code is very similar to the CFA model in Section 16.5, except for the inclusion of the causal structural components. To implement the causal structural components, we make use of another Mplus keyword "ON" to "regress on" these causal structural variables, which are bolded in the above Mplus code. SEM model fit is adequate with CFI and TLI values of 0.941 and 0.921, respectively, and an RMSEA of 0.045, which validates the SEM in Figure 20.1. The output for the full SEM is not shown here.

20.6.2 Analysis of NHANES Data Using SAS Program

The CALIS procedure in SAS/STAT is a general SEM tool. PROC CALIS uses complete data during the modeling. Thus, if a particular respondent skips survey items or did not complete their entire survey, then some data elements are missing. The PROC CALIS procedure will not include the respondent in the analysis.

SAS Program Code:

```
PROC CALIS DATA=CALIS NOBS=616 MODIFICATION; weight WTSA2YR;
FACTOR
  FTOBACCO  -> NSMOKE NCIGH,
  FVOC     -> XYLENE1S XYLENE2S ETHLBENS,
  FRF      -> NEVS PEFS FVCS;
PVAR
  FTOBACCO FVOC FRF =3*1;
COV
  FTOBACCO FVOC FRF =3*0;
  RUN;
```

NOBS allow the number of survey respondents in a dataset.
MODIFICATION" allows SAS to create a better fitted model; this allows one to get a series of possible modifications to obtain a better fit;.
FACTOR allows an outline of the factor –starting with F in front of factor, list all the variables associated with that factor.

PVAR specifies the parameters for factor variances and error variances of manifest variables.
COV specifies any suspected correlations between factors. The 3*1 indicates 3 factors with a variance of 1. This allows a fix on the variances in model identification. We can allow factors to be correlated. The 3*0 allows the correlation between factors to be zero, thereby allowing recommendations between factors.

SAS Output and Interpretations

The CALIS Procedure
Covariance Structure Analysis: Model and Initial Values
Modeling Information
Maximum Likelihood Estimation

Data Set	WORK.CALIS
N Records Read	616
N Records Used	616
N Obs	616
Model Type	FACTOR
Analysis	Covariances

	Variables in the Model
Variables	NSMOKE NCIGH NEVS PEFS FVCS ETHLBENS XYLENE1S XYLENE2S
Factors	FTOBACCO FVOC FRF
	Number of Variables = 8
	Number of Factors = 3

The basic modeling information using all 616 records. There are 8 variables and 3 factors.

The factor loading matrix is given. These bits of information allow the researcher to see a summary of things.

The CALIS Procedure
Covariance Structure Analysis: Descriptive Statistics

Simple Statistics

Variable	**Mean**	**Std Dev**
NSMOKE	0.17019	134.06289
NCIGH	1.58941	1603
NEVS	10.19225	566.33254
PEFS	10.29057	530.85818
FVCS	10.39048	551.59526
ETHLBENS	10.03541	526.44654
XYLENE1S	10.03935	566.26858
XYLENE2S	9.96530	415.82279

The mean and standard deviations of the observed variables are given. These bits of information allow the researcher to see a summary of observed variables.

The CALIS Procedure
Covariance Structure Analysis: Maximum Likelihood Estimation

	Fit Summary	
Modeling Info	Number of Observations	616
	Number of Variables	8
	Number of Moments	36
	Number of Parameters	16
	Number of Active Constraints	0
	Baseline Model Function Value	4.0217
	Baseline Model Chi-Square	2473.3519
	Baseline Model Chi-Square DF	28

	Pr > Baseline Model Chi-Square	<.0001
Absolute Index	Fit Function	0.1061
	Chi-Square	65.2436
	Chi-Square DF	20
	Pr > Chi-Square	<.0001
	Z-Test of Wilson & Hilferty	4.6885
	Hoelter Critical N	297
	Root Mean Square Residual (RMR)	33415.8127
	Standardized RMR (SRMR)	0.0623
	Goodness of Fit Index (GFI)	0.9756
Parsimony Index	Adjusted GFI (AGFI)	0.9560
	Parsimonious GFI	0.6968
	RMSEA Estimate	0.0606
	RMSEA Lower 90% Confidence Limit	0.0446
	RMSEA Upper 90% Confidence Limit	0.0774
	Probability of Close Fit	0.1307
	ECVI Estimate	0.1589
	ECVI Lower 90% Confidence Limit	0.1250
	ECVI Upper 90% Confidence Limit	0.2053
	Akaike Information Criterion	97.2436
	Bozdogan CAIC	184.0156
	Schwarz Bayesian Criterion	168.0156
	McDonald Centrality	0.9639
Incremental Index	Bentler Comparative Fit Index	0.9815
	Bentler-Bonett NFI	0.9736
	Bentler-Bonett Non-normed Index	0.9741
	Bollen Normed Index Rho1	0.9631
	Bollen Non-normed Index Delta2	0.9816
	James et al. Parsimonious NFI	0.6954

The Baseline Model Chi-Square estimate is less than 0.05. Additionally, the Chi-Square estimates for the Absolute Index are also less than 0.05. This indicates that the *a priori* model may not be the best fit for the survey response data. Under the Absolute Index, one sees that the Hoelter Critical N is greater than 200, indicating that the sample size for the model is adequate to assess the model fit.

The Goodness of Fit Index (GFI) and the Adjusted GFI (AGFI) are interpreted similar to an R^2 estimate, in that an estimate closer to one indicates a better fitting model. In this example, we see that the GFI is estimated at 0.9827 and the AGFI is estimated at 0.9688. The Goodness of Fit estimates here indicate a relatively well-fitted model. Under the Parsimony Index, PROC CALIS estimates Root Mean Square Error of Approximation (RMSEA) as 0.0606. Interpretation of this statistic is that the closer to zero, the better the model fit. Less than 0.05 is preferable, which the model meets. The Bentler-Bonett Normed Fit Index (NFI). The Incremental Index statistics test whether the fitted model is a better fit than an independence model. The preferable estimate is greater than 0.80, which this model achieves. Based on the model fit statistics, the researcher concludes that the *a priori* model may not be the best fitted model based on our data. Fortunately, through the PROC CALIS option "modification", SAS generates alternative factor assignments, covariance structures, and error variances and covariances that the researcher can consider when developing a better fitted model.

The CALIS Procedure
Covariance Structure Analysis: Maximum Likelihood Estimation

Rank Order of the 10 Largest LM Stat for Factor Loadings

Variable	Factor	LM Stat	Pr > ChiSq	Parm Change
ETHLBENS	FTOBACCO	11.13053	0.0008	74.38800
PEFS	FTOBACCO	2.66383	0.1027	-16.14883
ETHLBENS	FRF	2.21924	0.1363	34.28644
NSMOKE	FRF	1.55396	0.2126	-3.52761
XYLENE2S	FTOBACCO	1.45942	0.2270	14.74066
FVCS	FTOBACCO	0.57367	0.4488	7.79645
FVCS	FVOC	0.41649	0.5187	7.35992
NEVS	FVOC	0.38459	0.5352	-13.27389
NEVS	FTOBACCO	0.37493	0.5403	-11.82952
NSMOKE	FVOC	0.31132	0.5769	1.69472

When the survey researcher specifies the "modification" option within the PROC CALIS procedure, SAS generates a series of tables at the end of the SAS output, which provides recommended changes to the factor – variable relationships. Here is a list of data-driven recommendations for the assignment of individual survey items into different construct domains. The first of the recommended item assignments has significant impact on the performance of the overall model, given the predicted Chi Square impact p-values (PR>ChiSq) less than 0.05. This adjustment is made by altering the original PROC CALIS code to reassign ETHLBENS item to different concept domains/factors. So, survey item ETHLBENS is reassigned from the original code to factor Ftobacco. .

20.6.3 Analysis of NHANES Data Using R Program

The first R package is "sem", developed by John Fox and company (https://cran.r-project.org/web/packages/sem/). It fits general linear structural equation models (with observed and latent variables). Another R package for SEM is "lavaan" developed by Yves Rosseel and his group. The detailed descriptions of "lavaan" can be accessed at their "lavaan" webpage (http://lavaan.ugent.be) or R CRAN (https://cran.r-project.org/web/packages/lavaan/). You can use "lavaan" to estimate a large variety of multivariate statistical models, including path analysis, CFA, SEM, and growth curve models. We recommend that interested readers review Yves Rosseel's description of this package and R examples at www.jstatsoft.org/article/view/v048i02.

```
library(readxl)
CALIS<-read.csv(("Data file path/nhanes.csv"))
attach(CALIS)
library(lavaan)
## This is lavaan 0.5-23.1097
## lavaan is BETA software! Please report any bugs.
semmodel <- '
    # latent variables
      FTOBACCO=~NSMOKE+NCIGH
      FVOC=~XYLENE1S+XYLENE2S+ETHLBENS
      FRF=~NEVS+PEFS+FVCS'
fit = cfa(semmodel,data=CALIS,std.lv = TRUE,WLS.V=WTSA2YR)
```

The "Model" command is of course the most important command to specify the hypothesized SEM to be fitted. The latent variable "TOBACCO" is measured "**BY** NCIGH NSMOKE", the latent variable "VOC" is measured "**BY** XYLENE1S XYLENE2S ETHLBENS", and the latent variable "RF" is measured "**BY** PEFS NEVS FVCS". The "NSMOKE@0" is to set the error variance to be zero due to the estimated negative error variance in the initial SEM fitting. The observations are measured by WTSA2YR.

```
coef(fit)
##  FTOBACCO=~NSMOKE   FTOBACCO=~NCIGH     FVOC=~XYLENE1S
##             0.533             4.003              1.956
##    FVOC=~XYLENE2S    FVOC=~ETHLBENS          FRF=~NEVS
##             1.950             0.427              1.085
```

```
##          FRF=~PEFS            FRF=~FVCS      NSMOKE~~NSMOKE
##             1.837                1.913              -0.035
##        NCIGH~~NCIGH XYLENE1S~~XYLENE1S XYLENE2S~~XYLENE2S
##            11.882                0.750               0.613
## ETHLBENS~~ETHLBENS           NEVS~~NEVS          PEFS~~PEFS
##             3.444                3.003               0.618
##         FVCS~~FVCS      FTOBACCO~~FVOC    FTOBACCO~~FRF
##             0.298                0.084              -0.047
##         FVOC~~FRF
##             0.004
summary(fit,fit.measures = TRUE,standardized = TRUE)
## lavaan (0.5-23.1097) converged normally after 130 iterations
##
##    Number of observations                      616
##
##    Estimator                                    ML
##    Minimum Function Test Statistic          38.169
##    Degrees of freedom                           17
##    P-value (Chi-square)                      0.002
##
```

> The first goodness-of-fit statistic is the Chi-Square Test of Model Fit, calculated as the sample size minus 1, times the minimum fit function. The Chi-square statistic is 38.169 with 17 degrees of freedom and a p-value of 0.002, which indicates that the fit of the data to the hypothesized CFA model is not entirely adequate.

```
## Model test baseline model:
##
##    Minimum Function Test Statistic        2640.189
##    Degrees of freedom                           28
##    P-value                                   0.000
##
## User model versus baseline model:
##
##    Comparative Fit Index (CFI)               0.992
##    Tucker-Lewis Index (TLI)                  0.987
##
## Loglikelihood and Information Criteria:
##
```

```
##     Loglikelihood user model (H0)               -8902.271
##     Loglikelihood unrestricted model (H1)       -8883.187
##
##     Number of free parameters                          19
##     Akaike (AIC)                                 17842.542
##     Bayesian (BIC)                               17926.584
##     Sample-size adjusted Bayesian (BIC)          17866.263
##
## Root Mean Square Error of Approximation:
##
##     RMSEA                                           0.045
##     90 Percent Confidence Interval         (0.026   0.064)
##     P-value RMSEA <= 0.05                          0.641
##
## Standardized Root Mean Square Residual:
##
##     SRMR              0.038
```

> For the asthma data, the RMSEA is 0.043, CFI is 0.992, and TLI is 0.987, which indicate the CFA model in Figure 20.2 fits the data adequately, and therefore validates the three latent constructs from the eight observed variables.

```
## Parameter Estimates:
##
##  Information          Expected
##  Standard Errors            Standard
##
## Latent Variables:
##         Estimate  Std.Err  z-value  P(>|z|) Std.lv  Std.all
##  FTOBACCO =~
##    NSMOKE     0.533    0.092    5.824    0.000    0.533    1.067
##    NCIGH      4.003    0.703    5.690    0.000    4.003    0.758
##  FVOC =~
##    XYLENE1S  1.956    0.116   16.898    0.000    1.956    0.914
##    XYLENE2S  1.950    0.114   17.036    0.000    1.950    0.928
##    ETHLBENS  0.427    0.079    5.407    0.000    0.427    0.224
```

```
##   FRF =~
##    NEVS       1.085    0.078  13.821    0.000    1.085    0.531
##    PEFS       1.837    0.068  26.851    0.000    1.837    0.919
##    FVCS       1.913    0.067  28.601    0.000    1.913    0.962
##
## Covariances:
##        Estimate  Std.Err  z-value  P(>|z|)  Std.lv  Std.all
##   FTOBACCO ~~
##    FVOC       0.084    0.043    1.939    0.053    0.084    0.084
##    FRF       -0.047    0.040   -1.179    0.238   -0.047   -0.047
##   FVOC ~~
##    FRF  0.004    0.043    0.098    0.922    0.004    0.004
##
## Variances:
##             Estimate  Std.Err  z-value  P(>|z|)  Std.lv  Std.all
##   .NSMOKE    -0.035    0.097  -0.360    0.719   -0.035   -0.139
##   .NCIGH     11.882    5.487   2.166    0.030   11.882    0.426
##   .XYLENE1S   0.750    0.375   1.998    0.046    0.750    0.164
##   .XYLENE2S   0.613    0.372   1.646    0.100    0.613    0.139
##   .ETHLBENS   3.444    0.197  17.467    0.000    3.444    0.950
##   .NEVS       3.003    0.176  17.103    0.000    3.003    0.719
##   .PEFS       0.618    0.118   5.235    0.000    0.618    0.155
##   .FVCS       0.298    0.123   2.423    0.015    0.298    0.075
##    FTOBACCO   1.000                               1.000    1.000
##    FVOC       1.000                               1.000    1.000
##    FRF        1.000                               1.000    1.000
```

The first goodness-of-fit statistic is the Chi-Square Test of Model Fit, calculated as the sample size minus 1, times the minimum fit function. The Chi-square statistic is 38.521 with 18 degrees of freedom and a p-value of 0.002, suggesting it is not entirely adequate. This chi-square statistic can be misleading. Another GOF statistic is the comparative fit index (CFI; Bentler 1990). The CFI is typically normalized from zero to 1.00, with values close to 1.00 for well-fitting model. A revised cutoff value close to .95 has more recently been advised (Hu and Bentler 1999). Another is the Tucker-Lewis fit index (TLI; Tucker and Lewis 1973). Others are the root mean square error of approximation (RMSEA; Steiger and Lind 1980) and the standardized root mean square residual (SRMR), where the RMSEA <0.05 indicates good fit, and values as high as .08 represent reasonable errors of approximation in the population (Browne and Cudeck 1993).

20.7 Discussions

The survey researcher should use logic to determine whether data-driven recommendations are appropriate to the context of the survey tool and factor model. This chapter served as an introductory overview to SEM, which is common in public health research. We illustrated the SEM with a compiled data from NHANES on asthma research. SEM is a useful approach for modeling the multivariate associations between indoor air pollutants and RF. SEM models not only the associations between indoor air pollutants and RF but also controls for the relationships among different types of indoor air pollutants and their related covariates. As SEM accommodates latent variables, multiple correlated predictors and outcomes can be included in the measurement of tobacco smoke exposure, VOCs, and RF. This approach provides strong evidence for the presence or absence of associations between indoor pollutants and RF.

In general, SEM allows for the modeling of complex relationships between predictor and outcome variables, accommodates the inclusion of both latent and observed variables, and controls for relationships between multiple predictor variables. These strengths make SEM a robust analytic technique that can be used effectively in public health applications.

20.8 Exercises

Consider the following variables: three measures of lung function peak (expiratory flow [PEF0, NEV, and FVC), tobacco use by the respondent, and tobacco exposure in the respondent's home. In the 2011–2012 NHANES, metabolites for VOCs examined in the urine samples include the following: xylenes are 2-methylhippuric acid, 3-methipurc acid, and 4-methipurc acid, ethylbenzene is phenylglyoxylic acid, and toluene is N-acetyl-S-(benzyl)-L-cysteine. The covariates are age, poverty status, height, and weight. Use the three constructs tobacco, VOC, and RF to examine another theory consists of any of the variables in the full dataset.

References

Bentler, P. M. (1990). Comparative fit indexes in structural models. *Psychological Bulletin*, *107*, 238–246. http://dx.doi.org/10.1037/0033-2909.107.2.238

Bollen, K. A. (1989). *Structural Equations with Latent Variables*. New York: Wiley.
Browne, M. W. and Cudeck, R. (1993). Alternative ways of assessing model fit. In Bollen, K. A., Long J. S. (eds.), *Testing Structural Equation Models*, pp. 136–162. Newbury Park: Sage.
Byrne, B. M. (2012). *Structural Equation Modeling with Mplus – Basic Concepts, Applications, and Programming*. New York: Taylor & Francis.
Chen, D. G. and Yung Y. F. (2023). *Structural Equation Modeling using R/SAS: A Step-by-Step Approach with Real Data Analysis*. Boca Raton, FL: Chapman & Hall/CRC.
Hu, L.-T. and Bentler, P. M. (1999). Cutoff criteria for fit indexes in covariance structure analysis: Conventional criteria versus new alternatives. *Structural Equation Modelling*, *6*, 1–55.
Kline, R. (2011). *Principles and Practice of Structural Equation Modeling*, 3rd edn. New York: Guilford Press.
Nurmatov, U., Tagieva, N., Semple, S., Devereux, G., and Sheikh, A. (2013). Volatile organic compounds and risk of asthma and allergy: A systematic review and meta-analysis of observational and interventional studies. *Primary Care Respiratory Journal*, *22*, PS9–PS15.
Shook-Sa, B. E., Chen, D. G., and Zhou, H. (2017). Using structural equation modeling to assess the links between tobacco smoke exposure, volatile organic compounds, and respiratory function for adolescents aged 6 to 18 in the United States. *International Journal of Environmental Research in Public Health*, *14*(10), 1112. https://doi.org/10.3390/ijerph14101112
Stapleton, M., Howard-Thompson, A., George, C., Hoover, R. M., and Self, T. H. (2011). Smoking and asthma. *Journal of the American Board of Family Medicine*, 24, 313–322.
Steiger, J. H. and Lind, J. C. (1980). Statistically based tests for the number of common factors. Paper presented at the Psychometric Society Annual Meeting, Iowa City, IA, June 1980.
Tucker, L. R. and Lewis, C. (1973). A reliability coefficient for maximum likelihood factor analysis. *Psychometrika*, 38, 1–10.
Westland, J. C. (2015). *Structural Equation Modeling: From Paths to Networks*. New York: Springer.

21

Longitudinal Data Analysis and Latent Growth Curve Modeling

21.1 Research Interest/Questions

Each year, approximately 200,000 women in the United States are diagnosed with breast cancer (Jemal et al. 2005). More recent studies suggest similar trends, though updated figures may vary. This diagnosis, along with the subsequent treatment process, can be a significant source of stress for women on multiple levels (Blomberg et al. 2009). Among women who undergo surgery for breast cancer, the postsurgical period is often marked by heightened distress, and effective psychosocial adjustment is crucial for both quality of life and physical recovery (Blomberg et al. 2009). Research shows that depression and anxiety during recovery can lead to immunosuppressive and anti-inflammatory effects that may impede healing (Elenkov and Chrousos 2002).

Despite the importance of psychosocial recovery following breast cancer surgery, there is still a gap in the literature regarding the mood (specifically anxiety and depression) and social adjustment of these women during their first-year postsurgery (Shimozuma et al. 1999). More research is needed to understand the factors influencing mood and social adaptation during this recovery period, which could inform therapeutic interventions aimed at supporting these women (Shimozuma et al. 1999). For example, factors such as age and the type of breast cancer surgery performed have been found to impact postsurgical recovery outcomes (Weichman et al. 2015; Yu et al. 2018).

This chapter is then aimed to use real data to answer two key research questions: (1) How do mood and social adjustment change after breast cancer surgery? (2) How do time-invariant factors, such as age and treatment type (lumpectomy or mastectomy), influence these changes? By analyzing data from 405 Hong Kong Chinese women postsurgery, we will examine longitudinal patterns in psychological morbidity and social adjustment over

DOI: 10.1201/9781003596370-10

three time points (1, 4, and 8 months). Data from 405 Hong Kong Chinese women who underwent cancer surgery will be used to illustrate the longitudinal patterns of change in personality in psychological morbidity and social adjustment at 1, 4, and 8 months postsurgery. With the longitudinal data collected from postsurgery at 1, 4, and 8 months, we will review and discuss the appropriate techniques of longitudinal data analysis (or linear growth curve model) and multidomain latent growth curve modeling.

21.2 Breast Cancer Data with Measure Patterns of Change in Personality Assessment

This dataset examines changes in mood and social adjustment among Hong Kong Chinese women following breast cancer surgery. This specific population was chosen due to cultural and healthcare differences that may influence recovery patterns. Data were collected over three waves across an 8-month period from 405 women recruited from six government-funded hospitals in Hong Kong. Women with linguistic or intellectual difficulties, an active psychiatric diagnosis, or uncontrolled metastatic brain disease were excluded to ensure consistency in responses and focus on the target sample.

Eligibility was determined through a review of medical records, and eligible women were approached within 2–12 days after surgery. After informed consent, a baseline face-to-face interview was conducted, focusing on demographic details and initial morbidity measurements. Follow-up data collection included telephone interviews at the first month (Wave 1), fourth month (Wave 2), and eighth month (Wave 3) postsurgery, which measured morbidity and social adjustment. Detailed information on the sample, data collection methods, and sampling procedures are described in Byrne et al. (2008).

The measures are summarized as follows:

1. **Mood:** Mood is represented by anxiety and depression symptoms and was measured using the 12-item Chinese Health Questionnaire (CHQ) (Byrne et al. 2008). Women responded to the items of the CHQ on a 4-point Likert scale with anchor points ranging from 0 ("Not at all") to 3 ("Much more than usual") for those tapping symptoms and from 1 ("Much better than usual") to 4 ("Much worse than usual") for those tapping positive states. Items was summed up to form a composite score for the scale. Higher scores indicate poorer mood (i.e. more negative symptoms of anxiety and depression).

2. **Social Adjustment.** Social Adjustment was measured using the 33-item Chinese Social Adjustment Scale. Women were asked to indicate perceived change in closeness and comfort levels for a range of five types of social relationships and situations that include family relationships, body image, friend relationships, social activities, and sexuality. For example, one item from the family subscale asks, "Compared to before you had treatment for your illness, how close have you felt to your children?". All items on the scale are on a 5-point Likert scale with anchors ranging from 1 ("Much less than previously") to 5 ("A lot more than previously"). Higher scores on the scale indicate better social adjustment.
3. **Time.** Time was analyzed as a continuous variable across three waves: (a) Wave 1, (b) Wave 2, (c) Wave 3 representing first month, fourth month, and eighth month postsurgery, respectively.
4. **Time-Invariant Predictors.** Age of the women was measured at baseline and recoded into a dichotomous variable with women whose age is less than 51 years (reference group) compared to women of age 51 and above. The type of breast cancer surgery was coded as a dichotomous variable with two levels: (a) lumpectomy (reference group) and (b) mastectomy.

Data are stored as *"hkcancer.csv"*, which can be loaded into R as follows:

```
dCancer <- read.csv("hkcancer.csv", na.strings="*",header=T)
# Check the data dimension
> dim(dCancer)
[1] 405  10
> # Print the first 6 observations
> head(dCancer)
```

ID	MOOD1	MOOD4	MOOD8	SOCADJ1	SOCADJ4	SOCADJ8	Age	SurgTx	AgeGrp
1	15	NA	NA	95.906	NA	NA	70	Mastectomy	Older
2	16	25	22	114.889	105.111	90.444	47	Mastectomy	Younger
3	37	26	25	80.667	95.333	95.333	47	Mastectomy	Younger
4	19	16	15	112.839	108.581	99.000	52	Mastectomy	Older
5	13	16	14	115.000	105.000	101.000	43	Mastectomy	Younger
6	21	28	19	106.452	114.968	107.516	34	Lumpectomy	Younger

Table 21.1 shows the descriptive statistics of the data at wave 1 (same descriptive statistics can be done for other two waves). The women in this sample had a mean age of 51.59 (*SD* = 10.68). 51% of the women was below 51 in age while 49% was 51 and above in age. 80% of the women had a mastectomy while 20% had lumpectomy.

TABLE 21.1

Descriptive Statistics

Variables	*n* (%)	*M*	*SD*
Age at Wave 1	405 (100%)	51.59	10.68
<51	205 (51%)		
51 and above	200 (49%)		
Surgery types			
Lumpectomy	79 (20%)		
Mastectomy	326 (80%)		

21.3 Longitudinal Data Analysis for Mood and Social Adjustment

21.3.1 Analytical Approach

The current study used longitudinal models to examine the changes in mood and social adjustment of a sample of women recovering from breast cancer in the first 8 months after surgery. Longitudinal data analysis is a commonly termed in statistical modeling, but in social sciences and other field, longitudinal data analysis is also called linear growth curve modeling. Therefore, in this chapter, we will use longitudinal data analysis and linear growth curve modeling interchangeably.

In this analysis, time-invariant predictors, such as age and surgery type, were also used to account for the change in the longitudinal outcomes across time. Since repeated data were collected from each woman with more than two waves of data, linear growth curve model is a suitable analysis that can model change across time to explain changes at the within-individual level as well as the interindividual level (Singer and Willett 2003; Chen and Chen 2021). An ordinary least-square regression is unable to account for the repeated measurements of time anymore.

We analyzed one growth curve model for each of the two longitudinal outcomes separately: mood and social adjustment. Following the methodological recommendations of Chen and Chen (2021), each model was developed in similar steps. The first step began with an unconditional linear growth curve model (only time variable was added as a predictor) with random intercept linear mixed-effects model (Model A). The second step was a similar model but with the addition of a random slope linear mixed-effects model (Model B). To assess if the addition of a random slope made a difference to the model, information criteria (AIC, Akaike information criterion; BIC, Bayesian information criterion) were used to evaluate the relative fit across the two models (Model A vs B). Lower values on these AIC and BIC statistics indicate better model fit (Schwarz 1978). In addition, a maximum likelihood ratio χ^2 test was conducted to compare between Model A and Model B to test differences when a random slope is added with no change in fixed effects (Singer and Willett 2003). This was tested by comparing the difference in the "-2loglikehood" of Model A and B to a χ^2-distribution based on their differences in degrees of freedom.

The third (Model C) was to add two interaction terms (three-way or two-way interactions) between the time-invariant covariates (age of women and surgery type) with time. If any of the interaction terms was not significant, for the sake of model parsimony, the final model (Model D) will keep only the main effect of that time-invariant covariate (even if there was no statistically significant main effect found). This is because we wanted to control for both time-invariant covariates in order to answer our research questions.

21.3.2 Data Analysis of "Mood"

To fit the longitudinal linear growth curve models with the "MOOD", we first need to *reshape* the data from wide format (i.e., by *individual*) to long format (i.e., by *individual* and *time*). This can be done using R library *tidyverse* with the following R code chunk:

```
# Call tidyverse to reshape the data
library(tidyverse)
# Get the "Mood" data
dMOOD <- dCancer[,c("ID","MOOD1","MOOD4","MOOD8",
"AgeGrp","SurgTx")] %>%
# Reshape the data to long format
 gather("MOOD1","MOOD4","MOOD8",key="Var",
 value="MOOD", na.rm=TRUE) %>%
# Create the "Time" variable by "separate" the "Var"
  separate(Var, into=c("tmp","Time"), sep=4, convert=TRUE)
```

The modeling approach described in Section 21.3.1 can then be implemented in the following R code chunk:

```
# Call the R library "nlme"
library(nlme)
# MODEL A -> GROWTH MODEL (Random-Intercept)
modA <- lme(fixed = MOOD~Time, random = ~1|ID, data = dMOOD  )
summary(modA)
# MODEL B -> GROWTH MODEL (Random-Slope)
```

```
modB <- lme(fixed = MOOD~Time, random = ~Time|ID, data = dMOOD)
summary(modB)
# Model Selection using maximum likelihood ratio test
anova(update(modA), update(modB))
# MODEL C -> INTERACTION OF AGE & SURGERY TYPE
modC1 <- lme(fixed = MOOD~Time*SurgTx*AgeGrp,
random = ~1|ID, data = dMOOD)
summary( modC1)
# NO THREE-WAY INTERACTION, SO FIT 2-WAY INTERACTION MODEL
modC2 <- lme(fixed = MOOD~Time*SurgTx+Time*AgeGrp,
random = ~1|ID, data = dMOOD)
summary( modC2)
# MODEL D -> FIT MAIN MODEL
modD1 <- lme(fixed = MOOD~Time + AgeGrp + SurgTx ,
random = ~1|ID, data = dMOOD)
summary(modD1)
# SurgTx is not significant, so remove it and re-fit
modD2 <- lme(fixed = MOOD~Time + AgeGrp,
random = ~1|ID, data = dMOOD)
summary(modD2)
```

The above models (models A to D) were developed to examine the change in mood due to the effects of time while also controlling for two time-invariant predictors (age and surgery types). Model A showed an unconditional linear growth curve model fitted with a random intercept linear mixed-effects model. When a random slope was added in Model B, both AIC and BIC did not decrease significantly. For model selection, the results of the maximum likelihood ratio test indicated that adding a random slope made a trivial (i.e. statistically insignificant) difference 1.1287 on 2 degrees of freedom ($p = .5687$) as seen from the R output as follows:

```
> anova(update(modA), update(modB))
   Model df    AIC      BIC     logLik    Test  L.Ratio p-value
   modA   4 6482.409 6502.200 -3237.204
   modB   6 6485.280  6514.968 -3236.640 1vs2 1.128765  0.5687
```

Taking all this together, the results supported the conclusion that a random intercept model would be a more parsimonious model. Therefore, Models C and D were developed as random intercept models. In Model C, the three-way interaction of *Time*AgeGrp*SurgTx* was not statistically significant, leading to the reduced two-way interaction model of *Time*AgeGrp* + *Time*SurgTx*. The two-way interaction model was again not significant, leading to Model D1 of main effect model.

In Model D1, the main effect of *SurgTx* was not significant (*p*-value = .511) as seen from following R output:

```
Fixed effects:  MOOD ~ Time + AgeGrp + SurgTx
                     Value Std.Error  DF   t-value    p-value
(Intercept)       21.029843 0.6923276 658  30.375568  0.0000
Time              -0.190745 0.0440303 658  -4.332142  0.0000
AgeGrpYounger      1.910914 0.5478467 381   3.488045  0.0005
SurgTxMastectomy -0.444430 0.6755706 381  -0.657859  0.5110
```

The results in this Model D1 indicated that in the period of 8 months postsurgery, the mood of the women significantly improved ($b = -.191$, $p < .001$) after controlling for the effects of their age and surgery type. Main effect of surgery type was not statistically significant ($b = -.44$, $p = .51$) indicating that there was little or no average difference in mood between the group of women who had a lumpectomy and the group that had a mastectomy across the 8 months, which led to the final model, i.e., the Model D2, with *AgeGrp* only main effect model, where the estimated intercept is 20.676, slope of −0.191 and the parameter for the *AgeGrp* of 1.911 as seen from the following R output:

```
Random effects:
Formula: ~1 | ID
        (Intercept) Residual
StdDev:    4.719935 4.042909

Fixed effects:  MOOD ~ Time + AgeGrp
                  Value Std.Error  DF  t-value p-value
(Intercept)    20.676485 0.4364456 658 47.37472  0e+00
Time           -0.190617 0.0440282 658 -4.32943  0e+00
AgeGrpYounger   1.911355 0.5474601 382  3.49131  5e-04
```

This result indicated that those women who were 51 years and above were on average significantly better in their mood than those women below 51 years across the 8 months ($b = 1.91$, $p < .001$). This result can be graphically demonstrated in Figure 21.1.

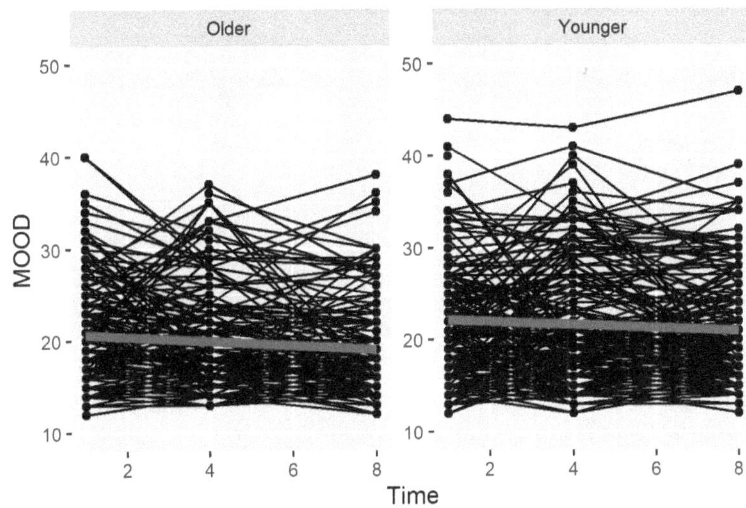

FIGURE 21.1
The Longitudinal Trend of "Mood" with Final Model Fit.

21.3.3 Data Analysis of "Social Adjustment"

The similar R code chunk as following to *reshape* the data from *wide format* to *long format* for longitudinal data analysis, which is named as "dSOCADJ":

```
# Get the "SOCADJ" data
dSOCADJ=dCancer[,c("ID","SOCADJ1","SOCADJ4","SOCADJ8",
        "AgeGrp","SurgTx")] %>%
# Reshape the data to long format
gather("SOCADJ1","SOCADJ4","SOCADJ8", key="Var",value="SOCADJ",
        na.rm=TRUE) %>%
# Create the "Time" variable by "separate" the "Var"
separate(Var, into=c("tmp","Time"), sep=6, convert=TRUE)
```

With this dataset, we can then implement the modeling approach described in Section 21.3.1. We fit the random intercept model in Model A and random slope model in Model B for model selection in the following R code chunk:

```
# MODEL A -> GROWTH MODEL (RANDOM INTERCEPT)
modA <- lme(fixed = SOCADJ~Time, random = ~1|ID,
data = dSOCADJ  )
summary(modA)
```

```
# MODEL B -> GROWTH MODEL (RANDOM SLOPE)
modB <- lme(fixed = SOCADJ~Time, random = ~Time|ID,
data = dSOCADJ)
summary(modB)
# Model Selection using anova
anova(update(modA), update(modB))
```

The associated result from this maximum likelihood ratio test is as follows:

```
Model df    AIC      BIC     logLik  Test  L.Ratio p-value
modA  4 7396.937 7416.767 -3694.468
modB  6 7392.174 7421.919 -3690.087 1 vs 2 8.762611  0.0125
```

It can be seen that with the random slope model in Model B in examining the change in social adjustment due to the effects of time, both the AIC and BIC decreased from the random intercept model in Model A. In addition, the likelihood ratio statistic is 8.762 with two degrees of freedom, which produced a *p*-value of .0125, and this is statistically significant. This result supported the decision that a random slope model would be a better model for the social adjustment longitudinal data to model the social adjustment trend by time.

Therefore, Models C and D were developed based on the random slope models. Starting with Model C1 with three-way interaction of Time*SurgTx*AgeGrp, it was found that the three-way interaction was not statistically significant (*p*-value of .5708). Consequently, the two-way interaction model (Model C2) for Time*SurgTx+Time*AgeGrp was fitted. We found that the interaction of Time*AgeGrp was not statistically significant (*p*-value of .4184). Hence, in the final Model D, we reduced the Model C2 by excluding the interactional term for age but retained the main effect of age for the sake of model parsimony and to retain the effect of age as Time*SurgTx+Time+AgeGrp, and we found again that the AgeGrp was not statistically significant. The final model in Model D2 is further reduced to exclude the AgeGrp. This modeling process are summarized in the following R code chunk:

```
# MODEL C -> Interaction with AGE & SURGERY TYPE
modC1 = lme(SOCADJ~Time*SurgTx*AgeGrp,random=~Time|ID,
data = dSOCADJ)
summary(modC1)
```

```
modC2 = lme(SOCADJ~Time*SurgTx+Time*AgeGrp,random=~Time|ID,
data = dSOCADJ)
summary(modC2)
# MODEL D -> FINAL MODEL
modD1 = lme(SOCADJ ~Time*SurgTx+Time+AgeGrp, random = ~Time|ID,
data = dSOCADJ)
summary(modD1)
modD2 <- lme(SOCADJ~Time*SurgTx, random = ~Time|ID,
data = dSOCADJ)
summary(modD2)
```

Part of the output of this Model D2 is as follows:

```
Random effects:
Formula: ~Time | ID
              StdDev   Corr
(Intercept) 8.691423 (Intr)
Time          0.596256 -0.486
Residual     5.489700

Fixed effects:  SOCADJ ~ Time * SurgTx
                          Value Std.Error  DF  t-value  p-value
(Intercept)           103.73834 1.1894177 665 87.21776  0.0000
Time                   -0.43878 0.1481460 665 -2.96182  0.0032
SurgTxMastectomy       -3.58092 1.3335998 384 -2.68515  0.0076
Time:SurgTxMastectomy   0.43055 0.1663302 665  2.58853  0.0098
```

Results from this Model D2 indicated that in the period of 8 months postsurgery, the interaction between the surgery type with the trend of the social adjustment of the women was statistically significant (slope of Time:SurgTxMastectomy 0.43055, $p = .0098$). This significant interactional effect of surgery type with time indicated that there was a significant difference in social adjustment between the group of women who had a lumpectomy and the group that had a mastectomy across the 8 months. Post hoc analysis indicated that the group of women who had a lumpectomy surgery had significantly deteriorated in the social adjustment across the 8 months (slope of −0.43878 with p-value of .0032) while the women who had a mastectomy surgery did not have any significant changes in social adjustment (slope of $-0.43878 + 0.43055 = -0.00823$, $p > .5$).

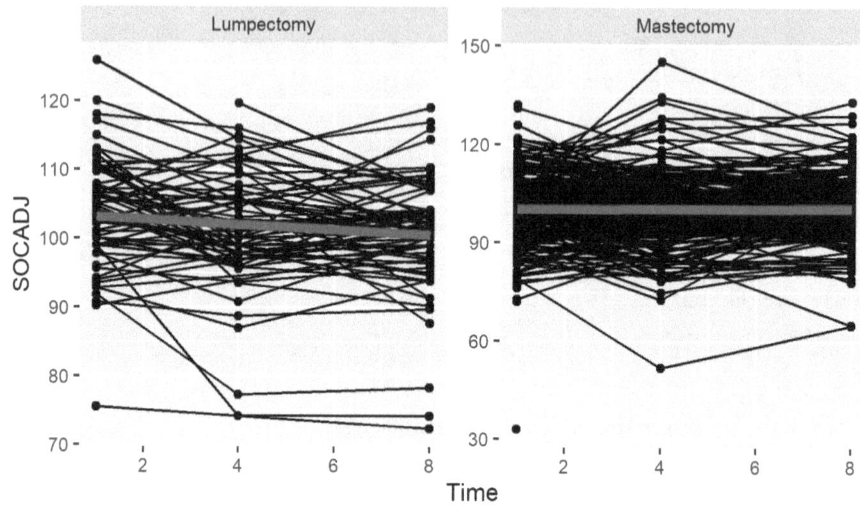

FIGURE 21.2
Changes in Social Adjustment for Surgery Type across 8 Months.

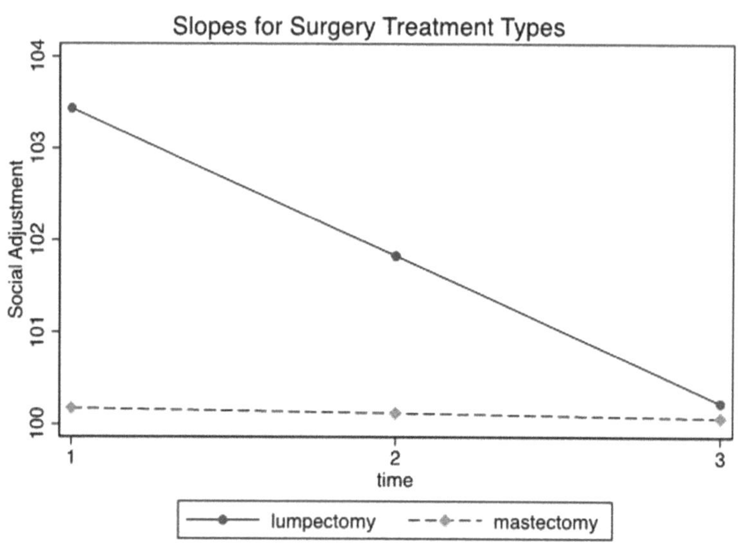

FIGURE 21.3
Mean Changes in Social Adjustment for Surgery Type across 8 Months.

This conclusion can be graphically displayed in Figure 21.2, which shows the changes in social adjustment scores for each surgery type along with the observed longitudinal social adjustment profiles for all women. The solid red lines in this figure are the estimated overall longitudinal growth trends for each surgery type, where it is evident that the significant decline in social adjustment in surgery type of lumpectomy, but less evident for mastectomy.

To better illustrate this trend, we summarized the mean social adjustment for all women by the three waves for the two surgery types. This is further illustrated in Figure 21.3. From this figure, it is more visible that the social adjustment for women with surgery type of lumpectomy is deteriorated faster than the social adjustment for all women with surgery type of mastectomy.

21.4 Joint Latent Class Growth Curve Modeling

Latent class growth curve (LGC) modeling is one of the most powerful models, which is typically included as part of the structural equation modeling (SEM), as described in Chapter 20, to analyze longitudinal data. It can also model longitudinal growth trajectories over time and test differences between groups (such as, in different intervention groups, age groups, etc) with both univariate and multivariate longitudinal outcomes.

As a general model for longitudinal growth, this model can incorporate both the within-individual and between-individual variations along with the multivariate longitudinal trajectories. Corresponding to the breast cancer data, for example, we can model the within-woman (i.e., within-individual) longitudinal change on both their mood (MOOD) and social adjustment (SOCADJ) over the 8-month period as well as the between-woman (i.e., between-individual) variations of their longitudinal trajectories. For detailed theoretical explanations of this joint LGC modeling, interested readers can refer to the Chapter 9 in Chen and Yung (2023), and this Section 21.4 is a continuation of that discussion focusing on R implementation.

In SEM, this latent class growth modeling is part of the popular R library *lavaan* (*la*tent *va*riable *an*alysis) with R function *growth*, which is freely available at www.lavaan.ugent.be/. We will review and illustrate how to use this library for this joint latent class growth modeling with the breast cancer data.

21.4.1 Joint LGC Modeling for "Mood" and "Social Adjustment"

As a powerful SEM technique, LGC modeling can easily incorporate multiple outcomes. In the Hong Kong women cancer data, two outcomes, "MOOD" and "SOCADJ", were measured at the 1-month, 4-month, and 8-month time points after surgery. We now illustrate the multiple-outcome LGC modeling by applying the LGC model depicted in Figure 21.4 to the cancer data.

We can observe the following features from Figure 21.4:

1. Modeling the LGC or trajectories for all outcomes simultaneously, as seen for both "MOOD" and "SOCADJ" together with ease and flexibility.

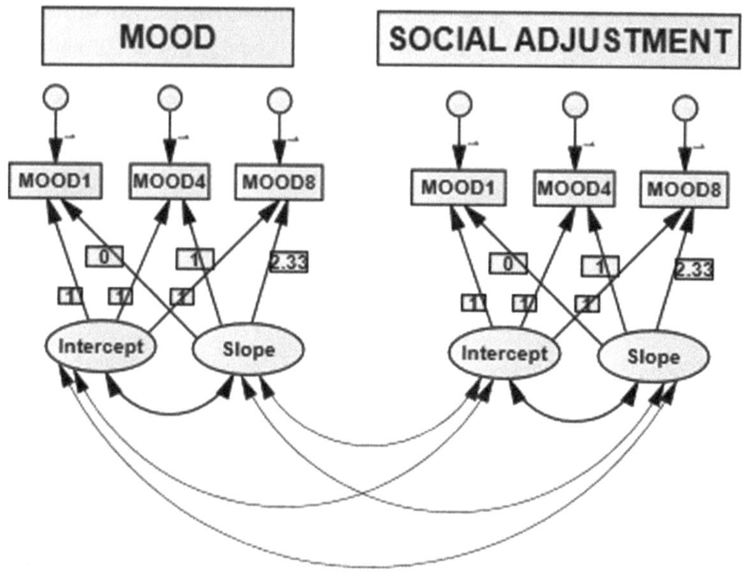

FIGURE 21.4
The LGC Model for Joint Modeling of "Mood" and "Social Adjustment".

2. Modeling the covariances (correlations) among the "latent" intercepts and slopes of all outcomes with the random intercepts and slopes between the two outcomes "MOOD" and "SOCADJ", as depicted by the four thin bidirectional paths connecting the random intercepts and slopes between the two outcomes.
3. Modeling the residual covariances (correlations) among the observed outcomes potentially by including bidirectional paths among the error terms of the observed "MOOD1", "MOOD4", "MOOD8", and "SOCADJ1", "SOCADJ4" and "SOCADJ8" to represent error covariances (correlations) if desired (not shown Figure 21.4 due to space limit in this book).

Detail model fitting and model selection can be found in Section 9.3 of Chen and Yung (2023), and the final model is presented here for further discussions. The following R code chunk is used to build the final model for this data:

```
# Joint LGC Model for Both MOOD and SOCADJ
mod4HK3 <- '
  # Intercept and Slope with MOOD
  iMOOD =~ 1*MOOD1 + 1*MOOD4 + 1*MOOD8
  sMOOD =~ 0*MOOD1 + 1*MOOD4 + 2.33*MOOD8
```

```
# Intercept and Slope with SOCADJ
iSOCADJ =~ 1*SOCADJ1 + 1*SOCADJ4 + 1*SOCADJ8
sSOCADJ =~ 0*SOCADJ1 + 1*SOCADJ4 + 2.33*SOCADJ8
# Add the error covariance based on modification indices
MOOD4 ~~ SOCADJ4
# Constrain the insignificant paths
# 1. covariances
iMOOD ~~ 0*sMOOD
iMOOD ~~ 0*sSOCADJ
sMOOD ~~ 0*iSOCADJ
# 2. Intercept
sSOCADJ ~ 0
# 3. Error variance
SOCADJ8 ~~ 0*SOCADJ8
'
```

Note in the above R code chunk, besides the regular random intercept and random slope modeling for both "Mood" and "SOCADJ", there are constraints for the three covariances with one of them for within-outcome, "iMOOD ~~ sMOOD" and the other two for between-outcome, "iMOOD ~~ sSOCADJ" and "sMOOD ~~ iSOCADJ". In addition, there is a constraint on intercept as "sSOCADJ ~ 0" and another for one error Variance as "SOCADJ8 ~~ 0*SOCADJ8" for model parsimony. With these constraints, the joint LGC model can be fitted as follows:

```
# Call growth to fit the model to the data
fitHK3 <- growth(mod4HK3,data =dCancer, estimator ="MLR", missing="ml" )
# Print the model fitting
summary(fitHK3,fit.measures = TRUE)
```

Calling R function "growth", we can fit the LGC model, the summary of the model fitting can be presented as follows (only relevant components are selected here with some explanations after #):

```
# The estimation:
  Estimator                                          ML
  Optimization method                            NLMINB
  Number of model parameters                         16
```

```
                                                   Used      Total
Number of observations                              386        405
Number of missing patterns                           15

# Model diagnostic statistics:
Model Test User Model:
                                               Standard     Scaled
  Test Statistic                                 13.756     10.488
  Degrees of freedom                                 11         11
  P-value (Chi-square)                            0.247      0.487
# The fitted model is satisfactory due to higher p-values here
Scaling correction factor                                    1.312
  Yuan-Bentler correction (Mplus variant)

# Satisfactory of CFI and TLI:
User Model versus Baseline Model:

  Comparative Fit Index (CFI)                     0.996      1.000
  Tucker-Lewis Index (TLI)                        0.995      1.001
  Robust Comparative Fit Index (CFI)              1.000
  Robust Tucker-Lewis Index (TLI)                 1.001

Loglikelihood and Information Criteria:
  Loglikelihood user model (H0)               -6852.169  -6852.169
  Scaling correction factor                       1.695
  for the MLR correction
  Loglikelihood unrestricted model (H1)        6845.291  -6845.291
  Scaling correction factor                       1.539
  for the MLR correction
  Akaike (AIC)                                13736.337  13736.337
  Bayesian (BIC)                              13799.631  13799.631
  Sample-size adjusted Bayesian (SABIC)       13748.865  13748.865

Root Mean Square Error of Approximation:
  RMSEA                                           0.025      0.000
  90 Percent confidence interval - lower          0.000      0.000
  90 Percent confidence interval - upper          0.062      0.047
  P-value H_0: RMSEA <= 0.050                     0.840      0.966
  P-value H_0: RMSEA >= 0.080                     0.004      0.000
  Robust RMSEA                                               0.000
  90 Percent confidence interval - lower                     0.000
  90 Percent confidence interval - upper                     0.065
  P-value H_0: Robust RMSEA <= 0.050                         0.855
  P-value H_0: Robust RMSEA >= 0.080                         0.012

Standardized Root Mean Square Residual:
  SRMR                                            0.035      0.035
```

```
Parameter Estimates:
  Latent Variables:
                     Estimate  Std.Err  z-value  P(>|z|)
  iMOOD =~
    MOOD1               1.000
    MOOD4               1.000
    MOOD8               1.000
  sMOOD =~
    MOOD1               0.000
    MOOD4               1.000
    MOOD8               2.330
  iSOCADJ =~
    SOCADJ1             1.000
    SOCADJ4             1.000
    SOCADJ8             1.000
  sSOCADJ =~
    SOCADJ1             0.000
    SOCADJ4             1.000
    SOCADJ8             2.330

 Covariances:
                     Estimate  Std.Err  z-value  P(>|z|)
  .MOOD4 ~~
    .SOCADJ4          -10.634    2.447    -4.346   0.000
  iMOOD ~~
    sMOOD               0.000
    sSOCADJ             0.000
  sMOOD ~~
    iSOCADJ             0.000
  iMOOD ~~
    iSOCADJ           -18.775   2.912  -6.447   0.000
  sMOOD ~~
    sSOCADJ            -0.853   0.499  -1.710   0.087
  iSOCADJ ~~
    sSOCADJ           -14.849  4.444  -3.341   0.001

 Intercepts:
                     Estimate  Std.Err  z-value  P(>|z|)
 sSOCADJ                0.000
 iMOOD                 21.405   0.313   68.351   0.000
 sMOOD                 -0.571   0.130   -4.398   0.000
 iSOCADJ              100.592   0.433  232.135   0.000
```

```
Variances:
                   Estimate  Std.Err  z-value  P(>|z|)
   .SOCADJ8           0.000
   .MOOD1            17.371    2.273    7.642    0.000
   .MOOD4            18.635    2.514    7.411    0.000
   .MOOD8             9.009    3.606    2.499    0.012
   .SOCADJ1          11.270    6.001    1.878    0.060
   .SOCADJ4          41.626    5.327    7.814    0.000
   iMOOD             22.007    2.719    8.093    0.000
   sMOOD              1.126    0.775    1.454    0.146
   iSOCADJ           81.989   13.035    6.290    0.000
   sSOCADJ           12.776    1.874    6.818    0.000
```

As seen from the model output, most parameter estimates are significant at the 0.05-level. In addition, the overall model fit measures is near perfect, as indicated by the "Comparative Fit Index (CFI)" = 1.00, the "Tucker-Lewis Index (TLI)" = 1.00, the "Root Mean Square Error of Approximation (RMSEA)" = 0.00 and the p-value for testing the RMSEA $\leq$ 0.05 (test of close-fit) is .966, and the "Standardized Root Mean Square Residual (SRMR)" = 0.035. With this satisfactory joint LGC model, we make some observations and conclusions as follows:

- Estimated covariances
  - Significant "MOOD4 ~~ SOCADJ4 = −10.634 (p-value < .01)", which indicates that the residuals of "MOOD" and "SOCADJ" at the fourth month correlates significantly. Hence, the correlation between "MOOD" and "SOCADJ" at the fourth month is not completely accounted for by the correlations between the random slopes and intercepts in the model.
  - Significant "iMOOD ~~ iSOCADJ = −18.775 (p-value = .000)", which means that the "Intercept" of "MOOD" and "Intercept" of "SOCADJ" are negatively correlated, indicating bad initial "MOOD" (higher mood values represent worse mood) at the first month was associated with lower level of initial social adjustment at the first month. Alternatively, the better the "MOOD" at 1 month, the better the "SOCADJ" at the beginning.
  - Significant "iSOCADJ ~~ sSOCADJ = −14.849 (p-value = .001)", which means that the "Intercept" of "SOCADJ" and "Slope" of "SOCADJ" are negatively correlated, indicating the better (or higher) the "SOCADJ at 1 month, the less the improvement. Alternatively, the worse (or lower) the "SOCADJ" at 1 month, the more improvement

of their "SOCADJ" during the 8-month postsurgery period.

- Estimated Intercepts or Means
  - Significant "iMOOD" and "iSOCADJ", which means that the means of "MOOD" and "SOCADJ", are significantly different from zero at the beginning of 1 month.
  - Significant "sMOOD = −0.571 (p-value = .000)". This negative significant negative mean slope for "MOOD" means that on average the "MOOD" of women postsurgery are improving during the 8-month postsurgery period.
  - Fixed zero mean for "sSOCADJ". A zero mean slope for "SOCADJ" means that on average the social adjustment of women are "not improving during the 8-month postsurgery period".
- Estimated "Variances"
- Significant error variances for "MOOD1", "MOOD4", "MOOD8", "SOCADJ1", and "SOCADJ4", which indicates that there are significant "residual" variations among the women on these measurements.
- Significant variances for "iMOOD", "iSOCADJ", and "sSOCADJ", which means that there are significant variations of initial "MOOD" and "SOCADJ" between women (i.e., as indicated by the variances of intercepts "iMOOD" and "iSOCADJ"). Also, the rate of improvement for "SOCADJ" between women are variable (i.e., as indicated by the variance of slope "sSOCADJ").
- Marginally significant variance for "sMOOD", which means that rate of improvement for "MOOD" might not vary among women. More evidence is needed to substantiate such an interindividual variation in improvement of mood.

There are many ways to explain these high variances. One of them is to include other covariates to explain these variations, which can be explored for the covariates of "AGE" groups and "Surgery" treatments in the following section.

21.4.2 Joint LGC Modeling for "Mood" and "Social Adjustment" with Covariates

The large variations might be attributed by different ages and treatment types among women. To explore and test this idea, the latent growth curve model can be adopted to include age groups and treatment types as covariates that predict the random intercepts and slopes for "MOOD" and "SOCADJ". The following *lavaan* R code shows the implementation:

```
mod4HK4 <- '
  # Intercept and Slope with MOOD
  iMOOD =~ 1*MOOD1 + 1*MOOD4 + 1*MOOD8
  sMOOD =~ 0*MOOD1 + 1*MOOD4 + 2.33*MOOD8
  # Intercept and Slope with SOCADJ
  iSOCADJ =~ 1*SOCADJ1 + 1*SOCADJ4 + 1*SOCADJ8
  sSOCADJ =~ 0*SOCADJ1 + 1*SOCADJ4 + 2.33*SOCADJ8
  # Add the error covariance based on modification indices
  MOOD4 ~~ SOCADJ4
  # Constrain the insignificant paths
  # 1. covariances
  iMOOD ~~ 0*sMOOD
  iMOOD ~~ 0*sSOCADJ
  sMOOD ~~ 0*iSOCADJ
  # 2. Intercept
  sMOOD ~ 0
  # 3. Error variance
  SOCADJ8 ~~ 0*SOCADJ8
  # Add regression paths to covariates
  iMOOD ~ AgeGrp + SurgTx
  sMOOD ~ AgeGrp + SurgTx
  iSOCADJ ~ AgeGrp + SurgTx
  sSOCADJ ~ AgeGrp + SurgTx
  '
# Call growth to fit the model to the data
fitHK4 <- growth(mod4HK4, data = dCancer,
estimator ="MLR", missing="ml" )
# Print the model fitting
summary(fitHK4,fit.measures = TRUE)
```

With the above R code chunk, the associated output for this model fitting can be shown as follows (again with explanations after ### and only kept the relevant output for brevity):

```
### The Estimation employed
  Estimator                                         ML
  Optimization method                           NLMINB
  Number of model parameters                        24
  Number of observations                           405
  Number of missing patterns                        16
```

```
### The model diagnostic statistics
Model Test User Model:
                                              Standard    Scaled
  Test Statistic                                15.147    12.532
  Degrees of freedom                                15        15
  P-value (Chi-square)                           0.441     0.638
  Scaling correction factor                      1.209
    Yuan-Bentler correction (Mplus variant)
  Comparative Fit Index (CFI)                    1.000     1.000
  Tucker-Lewis Index (TLI)                       1.000     1.007
  Robust Comparative Fit Index (CFI)            1.000
  Robust Tucker-Lewis Index (TLI)               1.006
Root Mean Square Error of Approximation:
  RMSEA                                          0.005     0.000
  90 Percent confidence interval - lower         0.000     0.000
  90 Percent confidence interval - upper         0.047     0.036
  P-value H_0: RMSEA <= 0.050                    0.965     0.994
  P-value H_0: RMSEA >= 0.080                    0.000     0.000
  Robust RMSEA                                             0.000
  90 Percent confidence interval - lower                   0.000
  90 Percent confidence interval - upper                   0.049
  P-value H_0: Robust RMSEA <= 0.050                       0.955
  P-value H_0: Robust RMSEA >= 0.080                       0.001
Standardized Root Mean Square Residual:
  SRMR                                           0.027     0.027

### Parameter Estimates:
  Standard errors                             Sandwich
  Information bread                           Observed
  Observed information based on                Hessian

Regressions:
                   Estimate  Std.Err  z-value  P(>|z|)
  iMOOD ~
    AgeGrp            1.696    0.582    2.915    0.004
    SurgTx            0.073    0.683    0.107    0.915
  sMOOD ~
    AgeGrp            0.152    0.207    0.736    0.462
    SurgTx           -0.434    0.173   -2.502    0.012
  iSOCADJ ~
    AgeGrp            0.960    0.987    0.972    0.331
    SurgTx           -3.337    1.057   -3.158    0.002
  sSOCADJ ~
    AgeGrp           -0.390    0.404   -0.966    0.334
    SurgTx            1.262    0.484    2.610    0.009
```

```
Covariances:
                   Estimate  Std.Err  z-value  P(>|z|)
  .MOOD4 ~~
    .SOCADJ4        -10.734    2.445   -4.391    0.000
  .iMOOD ~~
    .sMOOD            0.000
    .sSOCADJ          0.000
  .sMOOD ~~
    .iSOCADJ          0.000
  .iMOOD ~~
    .iSOCADJ        -19.282    2.896   -6.658    0.000
  .sMOOD ~~
    .sSOCADJ         -0.774    0.490   -1.579    0.114
  .iSOCADJ ~~
    .sSOCADJ        -14.192    4.361   -3.254    0.001

Intercepts:
                   Estimate  Std.Err  z-value  P(>|z|)
    .sMOOD            0.000
    .iMOOD           18.674    1.408   13.258    0.000
    .iSOCADJ        105.461    2.272   46.424    0.000
    .sSOCADJ         -1.944    1.123   -1.731    0.083

Variances:
                   Estimate  Std.Err  z-value  P(>|z|)
    .SOCADJ8          0.000
    .MOOD1           17.629    2.268    7.771    0.000
    .MOOD4           18.583    2.524    7.363    0.000
    .MOOD8            9.402    3.608    2.606    0.009
    .SOCADJ1         10.060    6.027    1.669    0.095
    .SOCADJ4         41.946    5.290    7.929    0.000
    .iMOOD           21.284    2.599    8.190    0.000
    .sMOOD            0.945    0.763    1.238    0.216
    .iSOCADJ         80.703   12.667    6.371    0.000
    .sSOCADJ         12.557    1.877    6.689    0.000
```

We can see that all overall model fit measures are nearly perfect: CFI and TLI are both about 1.00, RMSEA and SRMR are perfect, and the chi-square for model fit = 12.532 (df = 15, *p*-value = .6389). The test of close fit of RMSEA (smaller than 0.05) has a *p*-value of .994.

With this model fitting, we now investigate whether the large individual differences in random slopes and intercepts could be accounted for by the covariates "AgeGrp" and "SurgTx" in the model. It can be seen that the estimated residual variances for the "latent" factors of "iMOOD" (21.284), "iSOCADJ" (80.703) and "sSOCADJ" (12.557) in the current model are not much different from the estimated variances obtained from the model in Section 21.4.1, which is without the covariates "AgeGrp" and "SurgTx". This suggests that the age and the surgical type can only explain very small portions of variations of the random intercepts and slopes for "MOOD" and "SOCADJ". Or, individual differences in the random intercepts and slopes are only minimally accounted for by their age and surgical type. Further model investigations are needed to reduce these variance estimations. For example, other time-varying predictors at the within-individual level (such as personality traits, coping, and family relationships) as well as time invariant predictors at the interindividual level (such as socioeconomic status and hospital conditions) should be used in future models to improve the models' explanation for the changes in the outcomes.

Output from the corresponding "Regressions" component, we can make some conclusions as follows:

- For "MOOD":
  - "AgeGrp" has a significant effect on the initial "MOOD" (i.e., 1.696 with *p*-value of .004), which means that the initial "MOOD" at the first month was significantly lower (by 1.696 units) for women over the age of 50 (who had better initial mood) than that for younger women, keeping other factors constant.
  - "AgeGrp" does not have a significant effect on the "slope" for "MOOD, which means that the rate of changes in "MOOD" are not different for the two age groups.
  - "SurgTX" does not have a significant effect on the "Intercept" for "MOOD", which means that the initial "MOOD" are not different for the two types of surgery.
  - "SurgTX" has a significant effect on the "slope" for "MOOD". Women with mastectomy appeared to improve their mood faster than those with lumpectomy (by 0.434 unit, *p*-value = .012) for the 8-month period, keeping other factors constant.
- For "SocAdj":
  - AgeGrp does not have a significant effect on either the "Intercept" and "Slope" for "SOCADJ". This means that the two types of surgery appeared to have no effect on either the initial social adjustment or the rate of change across the 8-month period.
  - "SurgTX" has significant effects on both the "intercept" and "slope"

for "SOCADJ". Women with mastectomy had significantly lower "SOCADJ" at the beginning than those women with lumpectomy (by −3.337 units, p-value = .002). However, the rate of improvement in "SOCADJ" for women with mastectomy was faster (by 1.262 units, p-value = .009) for the 8-month period, keeping all other factors constant.

21.5 Conclusions and Discussions

The two primary aims of this chapter is to use latent class growth model to understand (1) how the mood and social adjustment of women change in the first 8 months after breast cancer surgery; (2) how time-invariant factors, such as age of the women and treatment types (lumpectomy or mastectomy), affect these changes in mood and social adjustment over the first 8 months. The results indicate that there are variations in how the outcomes changes over time and these changes can be explained by age of the women and the types of surgery that they had experienced.

As seen from the LGC, the mood of the women improved in the first 8 months after surgery. However, the social adjustment of the women deteriorated in the same period of time. One possible explanation for these differences is that social adjustment is also influenced by how others in the women's social environment (for e.g., family members, friends) perceive these women's condition and consequently relate to the women (Collins et al. 2004). For instance, if spouses have difficulties adapting to the women's postsurgery health conditions, marital interactions may be affected that could lead to poorer social adjustments among the women. This process may be independent of how women's mood is changing.

Also, interindividual differences, such as age differences, also account for the differences in mood scores across time. For instance, older women were more positive in mood than younger women. One explanation is that compared to younger women, older women may likely have experienced other physical losses or illnesses due to aging, and older women may likely have more life experiences that strengthen their coping skills. This may lead to older women having better emotional coping abilities that result in them having better mood outcomes than the younger women (Reuter et al. 2005). Changes in social adjustment scores over time were also explained by the types of surgery that the women experienced. Specifically, those women who had lumpectomy had significant deterioration in their social adjustment. Previous studies have argued that lumpectomy surgery, though a conservational treatment type, may result in poorer social adjustment because women feel that a more nonconservational treatment type such as mastectomy may be more effective (Pozo et al. 1992)

21.6 Exercises

1. Fit the first-order autoregressive joint LGC model for both the "MOOD" and "SOCADJ" data and test whether the autocorrelations are significant.
2. Fit the second-order autoregressive joint LGC model for both the "MOOD" and "SOCADJ" data, and test whether the autocorrelations are significant.

References

Blomberg, B. B., Alvarez, J. P., Diaz, A., Romero, M. G., Lechner, S. C., Carver, C. S., Holley, H., and Antoni, M. H. (2009). Psychosocial adaptation and cellular immunity in breast cancer patients in the weeks after surgery: An exploratory study. *Journal of Psychosomatic Research, 67*(5), 369–376. https://doi.org/10.1016/j.jpsychores.2009.05.016

Byrne, B. M., Lam, W. W. T., and Fielding, R. (2008). Measuring patterns of change in personality assessments: An annotated application of latent growth curve modeling. *Journal of Personality Assessment, 90*(6), 536–546. http://doi.org/10.1080/00223890802388350

Chen, D. G. and Chen, J. K. (2021). *Statistical Regression Modeling with R: Longitudinal and Miltu-level Modeling*. Cham, Switzerland: Springer.

Chen, D.G. and Yung, Y.-F. (2023). Structural equation modelling using R/SAS: A step-by-step approach with real data analysis. Boca Raton, FL: Chapman and Hall/CRC.

Collins, L. G., Nash, R., Round, T., and Newman, B. (2004). Perceptions of upper-body problems during recovery from breast cancer treatment. *Supportive Care in Cancer, 12*(2), 106–113. http://doi.org/10.1007/s00520-003-0554-5

Elenkov, I. J. and Chrousos, G. P. (2002). Stress hormones, proinflammatory and anti-inflammatory cytokines, and autoimmunity. *Annals of the New York Academy of Sciences, 966*(1), 290–303. http://doi.org/10.1111/j.1749-6632.2002.tb04229.x

Jemal, A., Murray, T., Ward, E., Samuels, A., Tiwari, R. C., Ghafoor, A., Feuer, E. J., and Thun, M. J. (2005). Cancer statistics, 2005. *CA: A Cancer Journal for Clinicians, 55*(1), 10–30. http://doi.org/10.3322/canjclin.55.1.10

Pozo, C., Carver, C. S., Noriega, V., Harris, S. D., Robinson, D. S., Ketcham, A. S., Legaspi, F. L., and Clark, K. C. (1992). Effects of mastectomy versus lumpectomy on emotional adjustment to breast cancer: a prospective study of the first year postsurgery. *Journal of Clinical Oncology, 10*(8), 1292–1298. http://doi.org/10.1200/JCO.1992.10.8.1292

Reuter, K., Classen, C. C., Roscoe, J. A., Morrow, G. R., Kirshner, J. J., Rosenbluth, R., Flynn, P. J., Shedlock, K., and Spiegel, D. (2005). Association of coping style, pain, age and depression with fatigue in women with primary breast cancer. *Psycho-Oncology, 15*(9), 772–779. http://doi.org/10.1002/pon.1012

Schwarz, G. (1978). Estimating the dimension of a model. *The Annals of Statistics, 6*(2), 461–464. http://doi.org/10.1214/aos/1176344136

Shimozuma, K., Ganz, P. A., Petersen, L., and Hirji, K. (1999). Quality of life in the first year after breast cancer surgery: rehabilitation needs and patterns of recovery. *Breast Cancer Research and Treatment, 56*(1), 45–57. http://doi.org/10.1023/A:1006214830854

Singer, J. D. and Willett, J. B. (2003). *Applied Longitudinal Data Analysis: Modeling Change and Event Occurrence*. New York: Oxford University Press. http://doi.org/10.1093/acprof:oso/9780195152968.001.0001

Weichman, K. E., Hamill, J. B., Kim, H. M., Chen, X., Wilkins, E. G., and Pusic, A. L. (2015). Understanding the recovery phase of breast reconstructions: Patient-reported outcomes correlated to the type and timing of reconstruction. *Journal of Plastic, Reconstructive and Aesthetic Surgery, 68*(10), 1370–1378. http://doi.org/https://doi.org/10.1016/j.bjps.2015.05.039

Yu, J., Son, W.-S., Lee, S. B., Chung, I. Y., Son, B. H., Ahn, S. H., Jo, M., and Lee, J. W. (2018). Uneven recovery patterns of compromised health-related quality of life (EQ-5D-3 L) domains for breast Cancer survivors: a comparative study. *Health and Quality of Life Outcomes, 16*. http://dx.doi.org/10.1186/s12955-018-0965-0

22

Latent Growth Mixture Joint Modeling in Intervention Research

22.1 Research Interest/Question

For intervention research in the social, psychological, behavioral, and health sciences, tracking time-varying phenomena is essential for assessing trajectories, changes, and growth patterns, particularly in studies examining intervention effectiveness in behavioral and health sciences. Examining change over time is particularly important for addressing complex questions about how specific interventions, programs, or policies influence socioemotional development, which can be measured through self-reports, behavioral assessments, and physiological indicator to ascertain whether their feelings, thought processes, and behaviors develop appropriately with time. When intervention researchers gather time-varying data, applying the appropriate statistical approaches is critical to optimally modeling change patterns and their antecedents as well as predicting their influence on other well-being and developmental outcomes. Typically, conventional growth curve modeling is used to estimate such trajectories as discussed in Chapters 12 and 21.

However, two potential challenges should be factored into data analytic strategies when dealing with longitudinal data: (a) the existence of heterogeneous trajectories among unobserved subpopulations and (b) the pervasiveness of outcome-dependent or nonignorable missing data, which is when missing values for a particular outcome variable is explained by unobserved values or unavailable data from another outcome variable being studied (Park et al. 2022). To ensure that such outcome-dependent missing data are handled correctly, it is crucial to apply appropriate modeling techniques to accommodate different types of missing data and prevent potential biases. This is partially addressed and illustrated in Chapter 16 for joint modeling with Bayesian approach.

To further address these modeling challenges effectively, this chapter advocates for the use of latent growth mixture joint modeling (LGMJM), a technique that combines elements of latent class analysis and joint modeling to enhance longitudinal data analysis. This more comprehensive modeling

DOI: 10.1201/9781003596370-11

approach relies on an integrative data harmonization (IDH) framework. Using the IDH framework to incorporate joint modeling techniques into the conventional latent class growth analysis enables researchers to efficiently assess heterogeneous subgroup trajectories while addressing challenges related to outcome-dependent missing data.

22.2 Motivation for a Latent Growth Mixture Joint Modeling Approach

22.2.1 Heterogeneous Trajectories among Unobserved Subpopulations

Subgroup identification and differentiation to determine intervention responsiveness are beyond typical growth curve models as discussed in Chapter 21. Despite being advanced, powerful, and valuable, this conventional growth curve modeling approaches rely on single estimates to approximate overall population trajectories (Jung and Wickrama 2008; Ram and Grimm 2009). When a large population contains distinct subgroups, this traditional approach fails to detect heterogeneity in trajectory patterns at the population level. Moreover, this approach assumes that independent variables and covariates are uniform in predicting growth patterns. Although multigroup growth curve modeling allows for assessing unique trajectories among subgroups, the determination of these subgroups is a priori.

By itself, conventional growth curve modeling does not reveal heterogeneous treatment effects. Instead, it treats all study participants as a homogenous sample by evaluating the intervention effect for the whole sample without assessing possible differential effects for subgroups (i.e., latent classes). However, there is evidence that heterogeneity in treatment effects can arise from various subgroups in intervention studies (Chung et al. 2021). Some subgroups may respond positively, while others react negatively to an intervention. Indeed, even in a failed intervention study, some subgroups of participants might benefit from the intervention.

Latent growth mixture modeling—a variant of growth mixture modeling—addresses these limitations by identifying and modeling unobserved subgroup trajectories (Ram and Grimm 2009). For example, studies on treatment adherence often use this method to categorize participants into groups based on their response patterns. By incorporating time-varying effects and trajectories, this strategy captures individual trajectories, growth patterns, and progression in ways that provide greater insights into often dynamic individual-level variation. Yet, latent growth mixture modeling also has a critical limitation: it does not efficiently address issues of outcome-dependent missingness (Park et al. 2022). Therefore, there is a need for an approach that uses all possible information to identify latent growth patterns within data collected from multiple sources. This chapter demonstrates how integrating joint modeling approaches as illustrated in Chapter 16 with latent growth mixture modeling as illustrated in Chapter 21 address this need together using the IDH framework.

22.2.2 Outcome-Dependent Missingness

In longitudinal studies, outcome-dependent missingness—i.e., when the probability of missing data depends on the value of the outcome variable—necessitates a joint modeling approach to overcome biases and invalidity statistical conclusions. In intervention research, tracking certain longitudinal data is only possible to the extent that participants remain in the study or are eligible for tracking. For instance, collecting academic performance data is only possible if students have not dropped out and remain in school. If a student's test score is missing because he dropped out of school at the time of the test, that missing test score is outcome-dependent and nonignorable. Likewise, child welfare researchers may track a child's socioemotional well-being outcomes until reunification with family, at which point, the child is no longer involved in the child welfare system. With these missing data examples, applying conventional growth modeling methods (e.g., latent class analysis, growth curve modeling, or growth mixture modeling) would imprecisely handle nonignorable missing values in outcome trajectories.

To overcome this challenge and obtain unbiased estimates, statistical models must account for time-to-event data, such as time to reunify with birth parents or time to school dropout (Chen et al. 2021; Brombin et al. 2016). In these cases, conventional statistical approaches cannot differentiate and address the existence of both valid missing and nonignorable missing values that may be present in the well-being or academic performance data. Statistical approaches such as sensitivity analysis, selection models, multiple imputation, and pattern mixture models are useful alternatives, but joint models have the capabilities to also harmonize longitudinal data (such as well-being or academic performance trends) and time-to-event data (such as reunification or school dropout status) to estimate the trajectories accurately. Additionally, joint models are particularly better at dealing comprehensive with complex nonignorable missingness patterns and flexibly estimating subject-specific trajectories, providing a higher level of inferential accuracy and robustness.

22.2.3 Latent Growth Mixture Joint Modeling

Latent Growth Mixture Joint Modeling (LGMJM) extends the capabilities of traditional latent class and latent class growth models (LGMM) by modeling an additional component when identifying latent classes or trajectories. This chapter is then used to advocate for increased LGMJM utilization to ensure a more efficient and unbiased concurrent estimation of time-to-event and longitudinal outcomes. Because longitudinal observations are time-dependent, they are measured with errors when participants drop out, but these dropouts are missing not at random (MNAR). LGMJM addresses this MNAR challenge by (a) linking the time-dependent longitudinal data (i.e., socioemotional scores in this dataset) to the time-to-event data (i.e., time-to-permanence in this dataset) and (b) simultaneously assessing the intervention effects on children's well-being trajectories and their likelihood of achieving permanence. This joint modeling method that links two data types also allows for properly handling commonly encountered outcome-dependent missing data using all available data.

The joint modeling framework is relatively novel to many social science disciplines, including the research communities in public health, social work, education, and political science. Studies from these disciplines scarcely use this framework when modeling trajectories, despite its ability to address complex research questions for which survival analysis and growth mixture modeling may concurrently be applied to address pervasive nonignorable missing data.

Because of its efficiency in addressing information loss when longitudinal and time-to-event data are analyzed together, a joint modeling approach properly accounts for outcome-dependent missingness. The approach can differentiate between outcome-dependent missingness and those that are not outcome-dependent and make the necessary statistical adjustments. This missingness differentiation ability and subsequent adjustment capabilities of the joint modeling approach when analyzing longitudinal data from randomized intervention studies have the advantage of preserving information from longitudinal and time-to-event data. By using all available data, LGMJMs can produce unbiased estimates and retain higher statistical power than the traditional approach (Park et al. 2022). The rest of the chapter is to use a real-world data to demonstrate the application and comparative advantage of LGMJM.

22.3 Real Data

22.3.1 Data Source

To demonstrate the LGMJM's application in real-world social intervention settings, we use data from the 5-year Illinois Birth through Three Title IV-E Waiver (IB3) demonstration project. In the IB3 demonstration study, the Illinois Department of Children and Family Services (IDCFS) randomly assigned child welfare-involved families to receive evidence-based parenting programs (i.e., either child–parent psychotherapy or the Nurturing Parenting Program) or services as usual between 2013 and 2017. The demonstration project hypothesized that evidence-based parenting interventions would foster steady nurturing parenting, promoting the socioemotional well-being of infants, toddlers, and preschoolers in legal custody of DCFS. The project also hypothesized that the parenting interventions would ultimately increase the likelihood of reunification with the family or alternative permanency options (Cooper et al. 2009; Lo et al. 2017).

22.3.2 Outcome Variables

The current study focuses on two outcome measures from the IB3 project: (a) family unification, measured as reunification with a birth parent or guardianship with a relative or nonrelative caregiver and (b) attachment, a longitudinal measure of the relationship between children (aged 3 years and younger) and significant adults, measured at multiple measurement occasions using the Devereux Early Childhood Assessment for Infants and Toddlers (DECA-I/T). These longitudinal data were collected from caregivers every 6 months, from when the child was 4 weeks old through 35 months. The current study uses information on all 1,597 children in the IB3 demonstration with up to three attachment scores. Higher attachment scores indicated a greater ability of the child to form and maintain positive relationships.

22.4 A Step-by-Step Demonstration Using R for Data Analyses

22.4.1 Analysis Framework

We juxtapose four modeling approaches: (a) separate analysis of longitudinal data only with growth curve modeling (i.e., Model 1 as discussed in Chapters 9 and 10); (b) separate analysis of survival data only with Cox regression (Cox 1972) (i.e., Model 2 as discussed in Chapter 14); (c) joint analysis of longitudinal and survival data with joint modeling framework without latent class identification (i.e., Model 3 as discussed in Chapter 21); and (d) joint analysis of longitudinal and survival data with LGMJM (i.e., Model 4 in this chapter). Following the final model determination, we perform additional analyses to understand the profile of each class trajectory identified through the joint modeling. In a separated study, Chen et al. (2021) demonstrated the application and merits of Models 1 to 3, using a small sample. This chapter builds on these three models by demonstrating the usability and added benefit of joint growth mixture modeling (i.e., Model 4).

We present and compare the results of the four models to highlight the benefits of the LGMJM over the classical longitudinal and survival models when dealing with outcome-dependent missingness and retention of both longitudinal and survival data. Using statistical notations, we describe the data format suitable for the analysis and each model's specification. We also provide a how-to example using the R statistical computing package "lcmm" (i.e., latent-class mixed model; Proust-Lima et al. 2017).

22.4.2 Separate Modeling of Longitudinal Data (Model 1)

First, we analyze the longitudinal attachment data by graphically investigating the relationship between the attachment scores and the number of days from case opening as shown in Figure 22.1. As seen from Figure 22.1, these DECA-I/T attachment scores behave in a complicated fashion: some children have increasing scores, some have decreasing scores, and others show more complex trends. However, because a global linear trend seemed reasonable, we used the linear mixed-effects model implemented in the R package "nlme" (nonlinear mixed-effects modeling).

To use "nlme", the longitudinal data should be in a person-time format with each participant and corresponding time in one row, as demonstrated in Table 22.1. Table 22.1 shows five participants with an ID of 1–5 (their real IDs are masked for privacy). Children 1 to 5 have 1, 2, 3, 4, and 5 observations, respectively. For each child, Table 22.1 shows their total days from screening

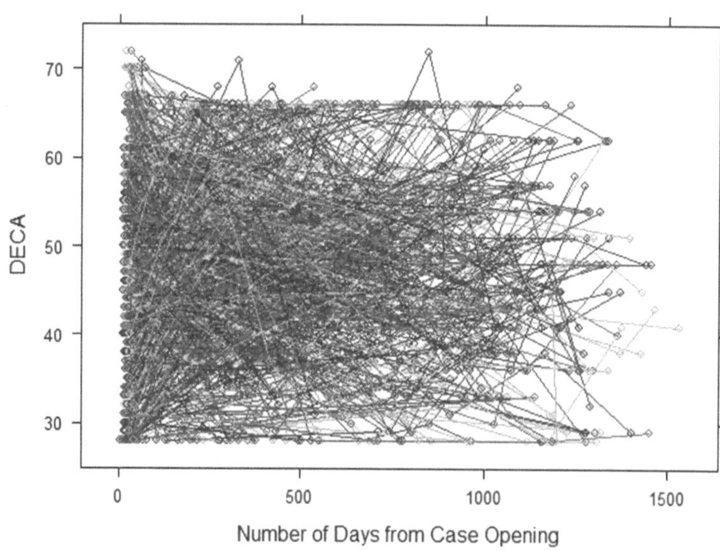

FIGURE 22.1
Longitudinal Trajectory Plots of Individual DECA Scores Observed at Different Time Points since Case Opening.

TABLE 22.1

A Snapshot of DECA Longitudinal Data for Five Participants in the Long Format Data

ID	Outcome	Days	Time in care	Treatment	DECA
1	0	36	483	Comparison	46
2	1	49	666	Intervention	44
2	1	252	666	Intervention	51
3	1	62	594	Intervention	71
3	1	67	594	Intervention	44
3	1	234	594	Intervention	47
4	1	62	594	Intervention	51
4	1	67	594	Intervention	54
4	1	67	594	Intervention	54
4	1	234	594	Intervention	36
5	0	26	1477	Comparison	48
5	0	152	1477	Comparison	56
5	0	564	1477	Comparison	47
5	0	741	1477	Comparison	50

Note: Days = total days at screen.

(i.e., "Days"), and their time in the foster care system (i.e., "TimeinCare") with their assigned treatment (i.e., "Treatment") and DECA-I/T attachment score on that day. This dataset is named "d2long" as reflected in the R code below.

We used the linear mixed effects (LME) modeling technique to statistically test the treatment effects and the temporal pattern (i.e., growth curve modeling). We started the modeling by testing the interaction between the treatment and the "Days," but the interaction term was not statistically significant. Therefore, we refit the model to estimate the main effect of "Treatment" and "Days" without the interaction term. We also tested whether a random slope LME fit better than the random intercept LME using the likelihood ratio chi-square test for model selection purposes.

The syntax used to fit the interaction models with random intercept LME in the R statistical programming language and computing environment with the "lme" function in R is

```
R> DECA.LME.Int1 = lme(DECA ~ Treatment*Days
            ,random = ~1|ID, data=d2long)
R> summary(DECA.LME.Int1)
```

As seen from the above R syntax, we call the R function "lme" (i.e., LMEs) to fit a mixed-effects model with *"DECA ~ Treatment*Days"* as the fixed-effects component and *"random = ~1 | ID"* as the random-effects component (i.e., random intercept). After fitting the model, we used the R function "*summary*" to print the model fit indices.

Similarly, the random slope LME for the interaction model with the R function "lme" and a change from a random intercept ("random=~1 | ID") to a random slope model ("random=~Days | ID") is as follows:

```
R> DECA.LME.Slope1 = lme(DECA ~ IB3*Days
            ,random=~Days|ID, data=d2long)
R> summary(DECA.Slope1)
```

The likelihood ratio chi-square test to compare the random intercept and random slope LME models can be implemented using the R function "anova" as follows:

```
R> anova(DECA.LME.Int1, DECA.LME.Slope1)
```

We found that the random slope LME (with AIC = 26,222.63, BIC = 26,272.15, and logLike = −13,103.32) had a better fit with the data, in comparison to the random intercept LME model (with Akaike information criterion (AIC) = 26,366.03, Bayesian information criterion (BIC) = 26,403.17, and logLike = −13,177.01). The resulting likelihood ratio chi-square statistic was statistically significant ($x^2 = 147.39, p < .0001$). Yet, even with this random slope LME model, the interaction term was not statistically significant ($\beta = 0.00058$, SE = 0.00118, $p = 0.6224$).

For model parsimony, we refit the models without the statistically nonsignificant interaction term by changing "*Treatment*Days*" to "*Treatment+Days*" as follows:

```
R> DECA.LME.Int2 = lme(DECA ~ Treatment+Days
            ,random = ~1|ID, data=d2long)
R> summary(DECA.LME.Int2)
R> DECA.LME.Slope2 = lme(DECA ~ IB3+Days
            ,random=~Days|ID, data=d2long)
R> summary(DECA.LME.Slope2)
R> anova(DECA.LME.Int2, DECA.LME.Slope2)
```

Similarly, we found that the random slope LME had a better model fit with the data, as the fit indices were lower (AIC = 26,209.23, BIC = 26,252.56, and logLike = −13,097.61) than those of the random intercept LME model fit indices (AIC = 26,352.77, BIC = 26,383.72, and logLike = −13,171.38): likelihood ratio x^2 = 147.538, p < .0001. As Figure 22.1 shows, with this random slope LME model, the DECA-I/T temporal trend was not statistically significant (β = 0.001, SE = 0.001, p = 0.1125). However, the IB3 treatment group experienced a statistically significant decrease in their DECA-I/T attachment scores (β = −1.256, SE = 0.412, p = .002).

22.4.3 Separate Modeling of Time-to-Event Data (Model 2)

Table 22.2 shows the data format for the separate survival analysis. In this table, "ID" represents the child identification number, which corresponds to "ID" in Table 22.1. "TimeInCare" is the time in foster care measured in months, and "Outcome" is a binary variable that indicates whether the child achieved family unification (1 = *yes* and 0 = *no*). "Treatment" is the treatment assignment for the child at the time of entry into foster care. In this R coding, we named the survival dataset "d2surv" as seen in the R code for the separate survival model.

This model can be implemented in R with the "survival" package (Therneau and Lumley 2009), as discussed in Chapter 14. The R code used for this separate Cox proportional hazards regression is as follows:

```
R> IB3.Cox = coxph(Surv(TimeInCare, Outcome)~Treatment
                  ,data = d2surv)
R> summary(IB3.Cox)
```

Using the time-to-event data and the Cox proportional hazard regression to identify intervention effectiveness revealed a statistically significant intervention effect (log hazard rate [HR] $beta_{IB3}$ = 0.197, *SE* = 1.217, p = .044) with a HR of 1.217, indicating an increased rate of family unification for the IB3 treatment group.

TABLE 22.2

A Snapshot of Survival Data for Five Participants in the Wide Format Data

ID	Treatment	Time in care	Outcome
1	Comparison	483	0
2	Intervention	666	1
3	Intervention	594	1
4	Intervention	594	1
5	Comparison	1477	0

22.4.4 Joint Modeling of Longitudinal and Time-to-Event Data (Model 3)

To quantify the effect of longitudinal DECA-I/T attachment scores on time to family unification, joint modeling allows researchers to associate the longitudinal outcome for each participant with their time-to-event data. The standard approach to achieving this type of joint modeling is to extend the hazard risk model in Cox regression models to incorporate the history of longitudinal data from the time of case opening to the time they achieve family unification (Therneau and Grambsch 2000), and this joining-modeling approach is discussed in Chapter 21.

In Chen et al. (2021), we illustrated the R package "JM" (i.e., joint modeling) developed by Rizopoulos (2010, 2012). However, this package cannot model latent classes. In this chapter, we use another R package, "lcmm" (i.e., latent class mixed modeling), which can be used to jointly model the longitudinal and time-to-event data while simultaneously estimating the latent classes (Proust-Lima et al. 2017). We highly recommend interested readers to follow their guidelines to analyze your own data. Readers may also follow our illustrative application for additional guidance.

With this R package, the joint modeling is a special case of the latent class joint modeling in which only one latent class exists. Using the "lcmm" package, the following R syntax can be used to fit the joint model for the DECA-I/T longitudinal data and permanency survival data:

```
R> lcjmod1 = Jointlcmm(fixed=DECA ~ Days+Treatment
                ,random=~Days,subject='ID'
                ,survival = Surv(TimeinCare,outcome)~Treatment
                ,hazard="3-quant-splines"
                ,hazardtype="PH",ng=1,data=d2long)
R> summary(lcjmod1)
```

Some explanations are given here. We call the function "Jointlcmm" for joint latent class modeling. For this function, we specify *"fixed=DECA ~ Days+Treatment"* as the fixed-effects component, similar to the "lme" in the separate longitudinal data analysis conducted in Model 1. We also specify *"random=~Days,subject='ID'"* as the random effects component with the random-slope model, as used in Model 1. The time-to-event data (i.e., family unification) are included as *"survival = Surv(TimeinCare, outcome)~Treatment"* with baseline hazard function piece-wise splines *"hazard="3-quant-splines"* and proportional hazard function *"hazardtype="PH"."* The number of latent classes is specified as *"ng=1,"* where "1" indicates the number of latent classes specified.

With this joint growth mixture modeling (JGMM) (ng = 1), the estimated longitudinal time effect ($\beta = 0.0009$, SE = 0.0006, $p = .119$) and the treatment effect ($\beta = -1.253$, SE $=0.412$, $p = .002$) are very similar to those shown in Model 1. The estimated intervention effect for the survival component is still significant ($\beta = 0.194$, SE $= 0.097$, $p = .046$). Compared to Model 2, the estimated SE is smaller (i.e., 0.097 in Model 3 vs. 1.217 in Model 2).

22.4.5 LGMJM of Longitudinal and Time-to-Event Data (Model 4)

To identify how many latent class trajectories fit the data best, we modeled a series of LGMJMs, starting with one latent class trajectory (i.e., ng = 1 in Model 3) and ending with four latent class trajectories (i.e., ng = 4). The results are summarized in Table 22.3. As seen from this table, the values of the likelihood function decrease from −16,851.76 for one latent class trajectory to −16,840.29 for two latent class trajectories, −16,825.65 for three latent class trajectories, and a slightly higher (−16831.45) for four latent class trajectories, which means that the model with three latent class trajectories fits better than the other models.

Using the likelihood ratio Chi-square test to select the best-fitting model among these four models, we found that the likelihood ratio Chi-square test between one- and two-class trajectories was statistically significant ($\chi^2 = 22.955$, df $= 4$, $p < .0001$), indicating the LGMJM with two-class trajectories has a statistically better fit than the one-class trajectory (Model 3). Similarly, the likelihood ratio chi-square test between two- and three-latent classes was statistically significant ($\chi^2 = 29.274$, df $= 4$, $p < .0001$), indicating the joint growth mixture model with three latent class trajectories has a statistically better fit than the two latent class trajectories. However, the likelihood ratio chi-square test between four- and three-class trajectories was not statistically significant ($\chi^2 = -11.596$, df $= 4$, $p = 1$). With the negative likelihood ratio statistic of −11.596, the LGMJM with four-class trajectories fits worse than the three-class trajectories. This means, of all the LGMJMs, the model with three-class trajectories fits the data best.

The distribution of the latent class trajectories (Table 22.3) further supports the choice of the three-class trajectories as the best LGMJM. The one-class

TABLE 22.3

Model Fit Indices and Percentage Breakdown of Tested Models

			Percentage distribution of the sample			
#Class	loglik	df	Class 1	Class 2	Class 3	Class 4
1	−16851.76	13	100.00	NA	NA	NA
2	−16840.29	17	2.13	97.87	NA	NA
3	**−16825.65**	**21**	**32.12**	**4.19**	**63.68**	**NA**
4	−16831.45	25	4.19	10.64	27.93	57.23

Note: loglik = likelihood ratio; df = degrees of freedom.

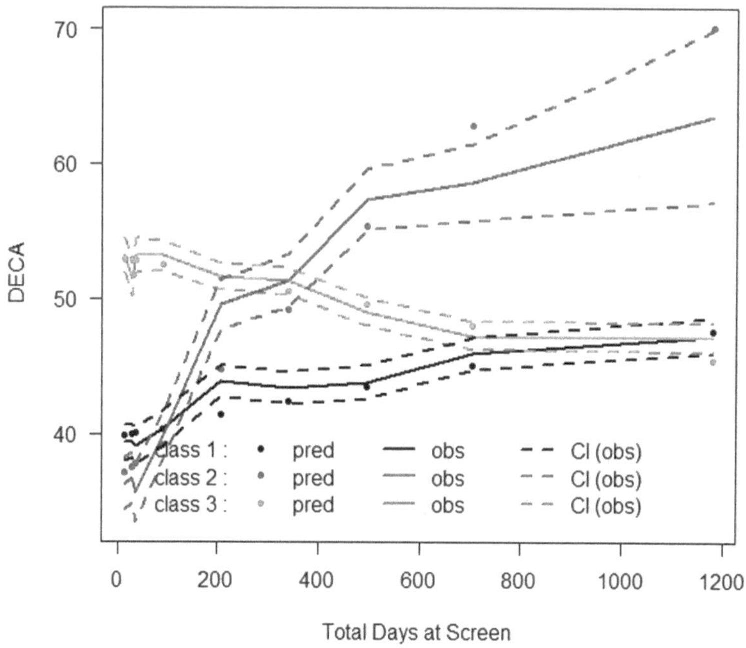

FIGURE 22.2
Three Latent-Class Trajectories Identified by the LCMJM.

trajectory lumps all 1,597 children into one class trajectory (i.e., Model 3), while the two-class trajectories have two trends with highly unbalanced sample distributions (i.e., 2.13% and 97.87%). However, the LGMJM with three latent class trajectories (Figure 22.2) identified three subgroups constituting 32.12% (class 1), 4.19% (class 2), and 63.68% (class 3) of the sample.

As shown in Figure 22.2, the children in *Class Trajectory 1* have a small but steady upward trajectory. They started with an average DECA-I/T score of 39, which increased to an average score of 44 by day 200, and slightly increased and stabilized at an average DECA-I/T score of 45. On the other hand, the children in *Class Trajectory 2* started very low with an average DECA-I/T score of 35, then increased substantially to an average score of 50 at 200 days in care, and kept increasing to an average score of 65 at the end of the study. The children in *Class Trajectory 3* had a declining trajectory, starting at an average DECA-I/T score of 53, then dropping to an average score of 46 at 700 days in care, and stabilizing afterward. The associated survival function for the probability of family unification for all three subgroups is shown in Figure 22.3. As seen in the figure, the children in class 2 have the shortest time achieving family unification, the children in class 1 have the longest, and the children in class 3 are in the middle.

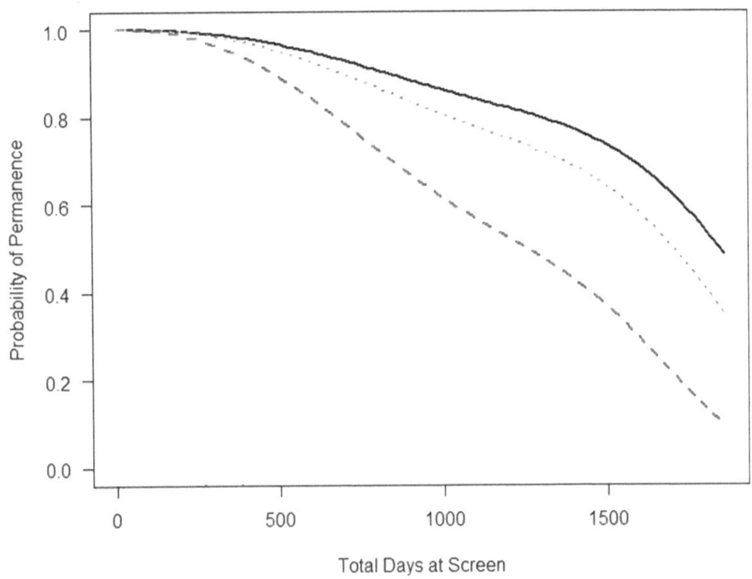

FIGURE 22.3
Estimated Class-Specific Survival Probability Function of Family Unification.

When we modeled the treatment variable as part of the final model (i.e., the three latent class trajectory model; ng = 3), the treatment effect was statistically significant ($\beta = -1.127$, SE = 0.410, $p = .006$). The estimated intervention effect for the survival component remains significant ($\beta = 0.191$, SE = 0.099, $p = .054$). Model 3 has a smaller SE than Model 2 (i.e., 0.097 and 1.217, respectively), indicating that some gains in predictive accuracy can be achieved by jointly analyzing both data types, as reported in Chen et al. (2021).

22.4.6 Profile Analysis of Latent Class Trajectories

One best practice after the conduct of person-centered analysis is a follow-up profiling and segmentation analysis that help create in-depth profiles for each identified latent class trajectory. Thus, analysts and researchers interested in applying LGMJM should consider conducting additional analysis to understand better the unique attributes of individuals within each class trajectory. This information provides a better understanding of complexity and variability within a population while also identifying risk or protective factors that inform policy decisions based on different trajectories of change over time. Such analysis can be bivariate analysis (e.g., chi-squares and t-tests) or multivariate analysis (e.g., logistic regression).

After using LGMJM to establish three-latent class trajectories, we further analyzed the latent class trajectories to highlight the typical characteristics

TABLE 22.4

The Number and Percentage of Children Who Achieved Family Unification by Latent Class Trajectory

	All Together			Comparison			Intervention		
Class trajectories	**No**	**Yes**	**%**	**No**	**Yes**	**%**	**No**	**Yes**	**%**
1. Steady upward trajectory	414	99	19.30	223	48	17.71	191	51	21.07
2. Sharp upward trajectory	30	37	55.22	18	12	40.00	12	25	67.57
3. Downward trajectory	729	288	28.32	384	145	27.41	345	143	29.30

Note: No = did not achieve family unification; yes = achieved family unification.

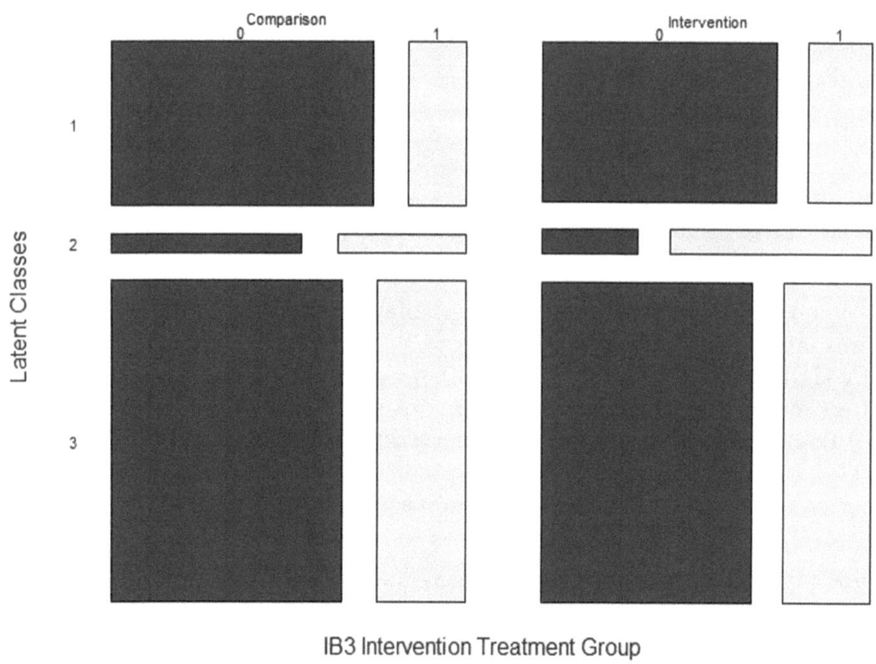

FIGURE 22.4
Cross-Classification of Children Who Achieved Family Unification for Three Identified Latent Class Trajectories.

of children and caregivers for each class trajectory. We conducted a series of chi-square tests, assessing the association between the three-class trajectories and several categorical variables. The first profile analysis assessed which of the three latent class trajectories was most likely to achieve family unification. Table 22.4 and Figure 22.4 present the latent class trajectories and the percentages of children in each who achieved family unification.

As seen in the full sample section of Table 22.4, 55.22% of children in class trajectory 2 (the sharpest upward trajectory group) achieved family unification compared to 19.30% in class trajectory 1 (the steady upward trajectory group) and 28.32% in class trajectory 3 (the downward trajectory group). The percentage difference is statistically significant ($\chi^2 = 43.71$, df = 2, $p < .0001$), meaning that the two subgroups whose trajectories of DECA-I/T attachment scores trended upward (i.e., trajectories 1 and 2 in Figure 22.2) also achieved more family unification than those children in the trajectory 3.

This three-part typology can be further segmented by the treatment variable to examine the intervention effect. As shown in Table 22.4, within trajectory 1 (i.e., the group that experienced a steady upward trajectory in attachment score), 21.07% of children in the treatment group achieved family unification compared to the 17.71% in the comparison group ($\chi^2 = 0.725$, df = 1, $p = .395$). Similarly, within trajectory 2 (i.e., the group that experienced the sharpest upward trajectory in attachment scores), 67.57% of children in the treatment group achieved family unification compared to the 40% in the comparison group ($\chi^2 = 4.038$, df = 1, $p = .044$).

More analysis was done. One notable finding concerns the association between latent class trajectories and age: more than 50% of younger children (i.e., ages below 35 months) were more likely to be in trajectory 3 (i.e., the group with a downward trajectory in attachment relationships) than the other classes. On the other hand, the majority (65.69%) of older children (i.e., at least 36 months old) belonged to trajectory 1, which experienced a steady but nondramatic increase in attachment scores.

22.5 Discussions and Conclusions

Collection of longitudinal data is very common in public health research, and so is the missing data phenomenon. When missing data are large and missing not completely at random because they are dependent on observed or unobserved data, then most of the current solutions (i.e., multiple imputations and full maximum likelihood) become less optimal (Bhattacharjee et al. 2020). LGMJM is well suited for such common scenarios. This chapter sought to create awareness and facilitate increased utilization of LGMJM, and we provided an overview and illustrated the application of the JGMJM using time-to-event and longitudinal data. We demonstrated that LGMJMs have three unique advantages over typically used modeling approaches.

First, we showed that JGMJMs produce a more efficient estimate of time-to-permanence and child–caregiver attachment trajectories than conventional approaches. By modeling the outcome variables concurrently within the same model (rather than sequentially), JGMJM can leverage shared error terms and avoid inflation of the error that arises from the accumulation of errors from two separate models that exceed what would be expected under the model assumptions of one statistical test. JGMJMs' smaller standard errors translate into increased precision. For instance, by extending the joint modeling to the typical growth mixture modeling, we cut the standard error by more than half from 1.217 to 0.097, helping narrow the range of values within which the estimated parameter can be explained.

Second, the LGMJM reduced biases in estimating both permanence and child–caregiver attachment trajectories compared to alternative models. As shown in this chapter, the JGMLM, unlike the alternative mixed-effect models and survival models, sufficiently addressed the outcome-dependent missingness, a capability that leads to greater statistical power and enhanced model performance. Fewer distortions in statistical estimation imply a closer approximation to the true population values. As statistical predictions become more accurate, reliable inferences and valid conclusions can be made, facilitating better and fairer policies and practices. This advantage of LGMJM is particularly important for studies with small sample sizes that may have little room to lose statistical power. For such small studies, the inability to efficiently deal with outcome-dependent missingness in some of the variables of interest could mean losing the statistical ability to detect an effect or a relationship if one exists.

Third, LGMJM can identify latent class trajectories that allow researchers to better understand which subgroups responded or did not respond to an intervention. This capability allows for a more nuanced insight into treatment effects by examining how study participants with varying class trajectories respond to a particular treatment. Such insights would enable better targeting of services and support to areas of need. Also, following the JGMM, analysts can perform secondary analysis to produce a profile of each identified latent class trajectory. Such person-centered profiling and segmentation capabilities can help identify differences in treatment response across different demographic, clinical, or other relevant subgroups. This information can guide personalized or targeted treatment approaches and improve patient outcomes.

These meaningful benefits of LGMJM improve the efficiency and accuracy of modeling in the social, psychological, and behavioral sciences. They are particularly beneficial for researchers seeking to unearth the presence of heterogeneous trajectories among unobserved subpopulations while accounting for outcome-dependent or nonignorable missing data. Its capacity to integrate multiple data types is particularly well suited for research in the public health sciences. This flexibility allows researchers to address diverse research questions in fields with complex, interrelated, and dynamic data. We encourage researchers who wish to identify and understand unobserved (i.e., latent) groups in a population to apply the syntax provided in this chapter to preserve information from longitudinal and time-to-event data, produce unbiased estimates, increase estimate precision, and retain higher statistical power than would be obtained via traditional approaches.

22.6 Exercises

Using the HIV/AIDS data, which is publicly available at the R packages "JM" and "JMBayes", the same analysis is performed in Section 22.4 from Model 1 to Model 4 and answered the following research questions:

1. How many latent classes can you identify?
2. What are the associated properties associated with each latent class?
3. Which model is better?

References

Bhattacharjee, A., Vishwakarma, G. K., and Banerjee, S. (2020). Joint modeling of longitudinal and time-to-event data with missing time-varying covariates in targeted therapy of oncology. *Communications in Statistics: Case Studies, Data Analysis and Applications*, *6*(3), 330–352. https://doi.org/10.1080/23737484.2020.1782286

Brombin, C., Di Serio, C., and Rancoita, P. M. (2016). Joint modeling of HIV data in multicenter observational studies: A comparison among different approaches. *Statistical Methods in Medical Research*, *25*(6), 2472–2487. https://doi.org/10.1177/0962280214526192

Chen, D. G., Ansong, D., Brevard, K. C., Okumu, M., and Bo, A. (2021). Joint modeling of longitudinal Devereux early childhood assessment data and time to permanency from randomized longitudinal intervention studies in social welfare research. *Journal of the Society for Social Work and Research*, 12(1): 247–262. www.journals.uchicago.edu/doi/10.1086/713768.

Chung, G., Ansong, D., Brevard, K., and Chen, D. G. (2021). Identifying treatment moderators of a trauma-informed parenting intervention with children in foster care: Using model-based recursive partitioning. *Child Abuse and Neglect, 117*. https://doi.org/10.1016/j.chiabu.2021.105065

Cooper, J. L., Masi, R., and Vick, J. (2009). *Social-emotional development in early childhood: What every policymaker should know*. National Center for Children in Poverty.

Cox, D. R. (1972). Regression models and life-tables. *Journal of the Royal Statistical Society, Series B*, *34*(2), 187–220. https://doi.org/10.1111/j.2517-6161.1972.tb00899.x

Jung, T. and Wickrama, K. A. (2008). An introduction to latent class growth analysis and growth mixture modeling. *Social and Personality Psychology Compass*, *2*(1), 302–317. https://doi.org/10.1111/j.1751-9004.2007.00054.x

Lo, S., Das, P., and Horton, R. (2017). A good start in life will ensure a sustainable future for all. *The Lancet*, *389*(10064), 8–9. https://doi.org/10.1016/S0140-6736(16)31774-3

Park, J. Y., Wall, M. M., Moustaki, I. and Grossman, A. H. (2022). A joint modeling approach for longitudinal outcomes and non-ignorable dropout under population heterogeneity in mental health studies. *Journal of Applied Statistics*, *49*(13), 3361–3376. https://doi.org/10.1080/02664763.2021.1945000

Proust-Lima, C., Philipps, V., and Liquet, B. (2017). Estimation of extended mixed models using latent classes and latent processes: The R Package lcmm. *Journal of Statistical Software*, *78*(2), 1–56. https://doi.org/10.18637/jss.v078.i02

Ram, N. and Grimm, K. J. (2009). Growth mixture modeling: A method for identifying differences in longitudinal change among unobserved groups. *International Journal of Behavioral Development*, *33*(6), 565–576. https://doi.org/10.1177/0165025409343765

Rizopoulos, D. (2010). JM: An R package for the joint modeling of longitudinal and time-to-event data. *Journal of Statistical Software*, *35*(9), 1–33. https://doi.org/10.18637/jss.v035.i09

Rizopoulos, D. (2012). *Joint Models for Longitudinal and Time-to-Event Data: With Applications in R*. Boca Raton, FL: Chapman and Hall/CRC Biostatistics Series.

Therneau, T. and Grambsch, P. (2000). *Modeling Survival Data: Extending the Cox Model*. New York: Springer-Verlag.

Therneau, T. and Lumley, T. (2009). "survival": Survival analysis including penalised likelihood. R Package Version 2.35-8. http://CRAN.R-project.org/package=survival

23

Causal Inference and Propensity Score Analysis

23.1 Research Interest/Questions

Causal inference for causal effect is the fundamental objective for research across disciplines. Although many statistical models, such as regression models presented in earlier chapters, aim to provide causal relationships among variables of interest, few can really offer estimates for causal effects due to confounding from observable and mostly unobservable factors. Unless such factors are all identified and controlled for, the statistical analyses can only be used to estimate the association, not the causation.

Consider this scenario, a researcher wants to evaluate the causal effect of a new health intervention to improve mental health recovery. If all patients in the new intervention group are from better healthcare facilities and the patients for the control group are from less advantaged healthcare facilities, one may see a difference in recovery rate between the two groups, with a higher rate of recovery from the new intervention group. However, this difference cannot be attributed to the causal effect of the new intervention. One cannot conclude that the new intervention treatment is more effective since the observed improvement could simply be due to the confounding factor of sampling. This confounding factor is, in fact, observed due to the sampling process, and the bias due to this confounding factor can be easily controlled for in regression models for causal effects. However, in real research and applications, such as in most observational studies, it is very difficult to observe all confounding factors, which are the sources leading to the so-called *hidden bias* in causal inference. Extensive discussions of this topic can be found in Rubin (1974, 1976, 2004, 2005) and He et al. (2016).

The propensity score (PS) analysis is widely used to adjust confounding effects in causal inference to compare treatment effects. It summarizes the covariate information into a single score. The PS is simply defined as the conditional probability of a treatment given patients' covariates (i.e., the probability of treatment assignment conditioning on observed confounding factors). Then, the PS is considered as a balanced score for the treatment

DOI: 10.1201/9781003596370-12

conditions. In this sense, PS analysis can be used to eliminate imbalances in baseline covariate distributions between treatments to estimate marginal treatment effects. Of course, PS is generally unknown and must be estimated, which is typically done by logistic regression.

The literature on PS analysis is rich with extensive publications and books. For additional references, we refer readers to Pan and Bai (2015) and Guo and Fraser (2015). In this chapter, we give an introductory overview of causal inference and discuss how to use PS analysis for causal inference with SAS and R.

23.2 German Breast Cancer Data

In this chapter, we revisit the German Breast Cancer data (Royston and Sauerbrei 2008, pp. 262–263; see http://biostat.mc.vanderbilt.edu/wiki/Main/GermanBreastCa) from Chapter 3. We use the 686 observations in this dataset to evaluate the effect of a medication, tamoxifen (TMX), on the response variable tumor size ("size"). It is a well-known fact that TMX is widely used for the treatment of both early and advanced estrogen receptor (ER)-positive (ER+) breast cancer in pre-/postmenopausal ("meno") women. It was approved by the U.S. Food and Drug Administration (FDA 2020) for the prevention of breast cancer in women at high risk of developing the disease. It is also approved for the reduction of contralateral (in the opposite breast) cancer. Two other risk variables used in this analysis are patients' age ("age") and patients' nodes number ("nodes"). The research question is then "Does tamoxifen affect tumor size in breast cancer?" for Hormone *Treatment* (TMX or no TMX) on *Outcome* (Tumor size) with adjustment of *Confounders* (age menopause and nodes).

23.3 Propensity Score Analysis

The PS is the probability (from 0 to 1) of an outcome belonging to a particular treatment group based on a given set of observed covariates. It is natural to use logistic regression with the group (treatment/control), a binary variable as the response, and the other covariates as the input variables.

From the fundamental level, PS analysis can only be conducted with two groups. If we have more than two groups, we have to conduct PS analysis two by two or use advanced PS analysis models. The method's performance is limited by how good the predictors used to generate it are. It eliminates cases with any missing data. It is only interested in the predicted probabilities in the logistic regression model.

With PS analysis, the PS is a balancing score between the two groups. The differences in the covariates between groups reduce into a single score, so if two groups are balanced on the PS, then they are balanced on all the observed covariates. There are four basic applications:

1. **Propensity score matching**: Match one or more control cases with a PS that is almost or equal to the PS for each treatment case.
2. **Stratification:** Divide the sample into strata based on rank-ordered PSs. Then compare groups within each stratum.
3. **Regression adjustment:** Include PSs as a covariate in a regression model used to estimate the treatment effect.
4. **Weighting:** Inverse probability of treatment weighting weights cases based on the inverse of the PS. This is similar to the use of survey sampling weights to ensure samples are representative of certain populations (Austin 2011).

23.4 Propensity Score Analysis Using SAS and R

23.4.1 SAS Program for Propensity Score

In SAS, we use PROC LOGISTIC or PROC GENMOD to obtain PSs. These scores are obtained because it is the conditional probability of each patient receiving a particular treatment based on pre-treatment variables. It relies on the predicted probabilities as a variable from fitting a logistic regression model with the following SAS program:

```
libname Research 'Data file path';
run;
data chap23;
set Research.propensity;
run;
proc logistic data = chap23;
class meno;
model hormon (event="had tamoxifen") = age nodes meno/link=logit rsquare;
output out = psds  pred = ps
xbeta=logit_ps;
run;
```

Estimate the PS: As in Chapter 8, fit the logistic regression model to obtain the predicted values and saved in the dataset called PSDS. We estimate the PS through the logistic regression model.

The LOGISTIC Procedure

Model Information	
Data Set	WORK.CHAP23
Response Variable	hormone
Number of Response Levels	2
Model	binary logit
Optimization Technique	Fisher's scoring
Number of Observations Read	686
Number of Observations Used	686

Analysis of Maximum Likelihood Estimates

Parameter		**DF**	**Estimate**	**Standard Error**	**Wald Chi-Square**	**Pr > ChiSq**
Intercept		1	-2.2028	0.6886	10.2336	0.0014
age		1	0.0279	0.0131	4.5618	0.0327
nodes		1	0.00295	0.0151	0.0381	0.8453
meno	Postmenopausal	1	0.4106	0.1332	9.5011	0.0021

We provide selected portions of the output as they pertain to the analysis. As seen from the output, the logistic regression with "Hormone" as the outcome variable and "age", "meno", and "nodes" as the independent variables. With the estimated PS, we can examine the PS distributions for each treatment group (i.e., with and without TMX).

```
PROC UNIVARIATE DATA=PSDS PLOT;
TITLE 'HISTOGRAMS OF PROPENSITY SCORES BY TREATMENT GROUP';
VAR PS;
CLASS HORMON;
HISTOGRAM PS / CTEXT=PURPLE CFILL=BLUE
KERNEL (K=NORMAL COLOR=GREEN W=3 L=1)
NORMAL (COLOR = RED W=3 L= 2)
NCOLS=1 NROWS=2;
INSET N='N' (COMMA6.0) MEAN='MEAN' (6.2)
MEDIAN='MEDIAN' (6.2) MODE='MODE'(6.2)
NORMAL KERNEL(TYPE) / POSITION=NW;
RUN;
```

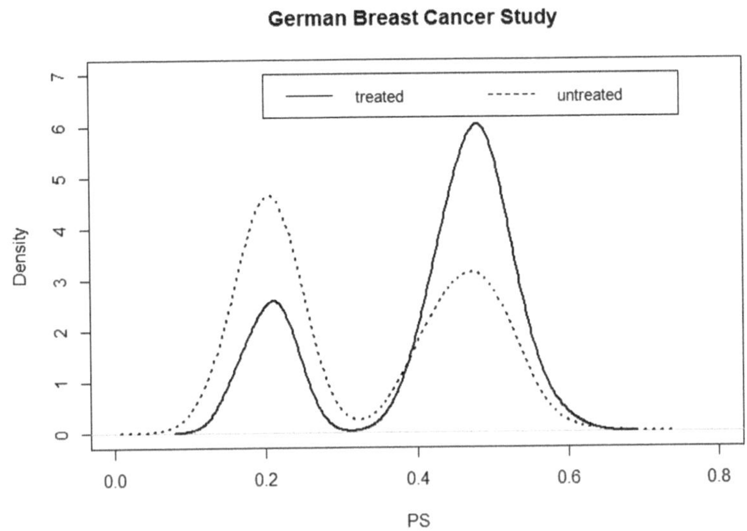

FIGURE 23.1
Distributions of PS by Both Treatments.

PS Balance: Based on the estimated PS, we balance the covariate imbalance distributions between treatment groups embedded in the observational study. Stratification and matching by PS are the two most commonly used methods to eliminate these imbalances to mimic data situation as in randomized control trials (RCTs). In PS stratification, patients are grouped in the same strata with similar or equal values for their propensity scores.

```
proc logistic data=chap23;
class meno;
 model hormon (event="had tamoxifen") = age nodes meno/link=logit
 rsquare;
output out = psds pred = ps
xbeta=logit_ps;
run;
proc rank data = psds groups=5
out = rank_ds;
ranks rank;
var ps;
data quintile;
```

```
set rank_ds;
quintile = rank + 1;
run;
proc glm data = quintile;
class hormon;
model size = hormon quintile /solution;
lsmeans hormon / om adjust = tukey pdiff cl;
quit;
```

Estimation of Treatment Effect based on PS: The ultimate goal in PS analysis is to estimate the treatment effect with the PS. For the German breast cancer data, we want to estimate the medication (i.e., Hormone) effect based on the PS.

The GLM Procedure

Source	DF	Type III SS	Mean Square	F Value	Pr > F
hormon	1	0.23401872	0.23401872	0.01	0.9296
quintile	1	97.92819341	97.92819341	3.27	0.0709

Parameter	Estimate		Standard Error	t Value	Pr > |t|
Intercept	4.190684887	B	0.49090201	8.54	<.0001
hormon had tamoxifen	-0.040078310	B	0.45315316	-0.09	0.9296
hormon no tamoxifen	0.000000000	B	.	.	.
Quintile	0.278234159		0.15378624	1.81	0.0709

The GLM Procedure

Source	DF	Type III SS	Mean Square	F Value	Pr > F
hormon	1	46.0855603	46.0855603	0.23	0.6353
quintile	1	124.0100008	124.0100008	0.61	0.4366

Parameter	Estimate		Standard Error	t Value	Pr > |t|
Intercept	30.46952409	B	1.28390721	23.73	<.0001
hormon had tamoxifen	-0.56242749	B	1.18517869	-0.47	0.6353
hormon no tamoxifen	0.00000000	B	.	.	.
quintile	-0.31310134		0.40221319	-0.78	0.4366

> Results from all these analyses indicate that the medication effect (i.e. Hormone effect) on breast cancer tumor size is not statistically significant, p=0.6353.

The GLM Procedure

Source	DF	Type III SS	Mean Square	F Value	Pr > F
hormon	1	46.0855603	46.0855603	0.23	0.6353
quintile	1	124.0100008	124.0100008	0.61	0.4366

Parameter	Estimate		Standard Error	t Value	Pr > |t|
Intercept	30.46952409	B	1.28390721	23.73	<.0001
hormon had tamoxifen	-0.56242749	B	1.18517869	-0.47	0.6353
hormon no tamoxifen	0.00000000	B	.	.	.
quintile	-0.31310134		0.40221319	-0.78	0.4366

> Results from all these analyses indicate that the medication effect (i.e. Hormone effect) on breast cancer tumor size is not statistically significant, p=0.6353.

The GLM Procedure
Least Squares Means
Adjustment for Multiple Comparisons: Tukey-Kramer

hormon	size LSMEAN	H0:LSMean1=LSMean2	
			Pr > |t|
had tamoxifen	28.9687054		0.6353
no tamoxifen	29.5311329		
hormon	size LSMEAN	95% Confidence Limits	
had tamoxifen	28.968705	27.130836	30.806574
no tamoxifen	29.531133	28.172315	30.889951

Least Squares Means for Effect hormon

i	j	Difference Between Means	Simultaneous 95% Confidence Limits for LSMean(i)-LSMean(j)	
1	2	-0.562427	-2.889459	1.764604

23.4.2 R Program for Propensity Scores

There are many R packages developed for propensity scoring analysis, such as "nonrandom", "Matching", and "MatchIt", to name a few. Susanne Stampf developed the package "nonrandom". This package offers a comprehensive data analysis when stratification and matching by the PS are undertaken. Several functions are implemented, including the selection of the PS model and estimating PS based on treatment or exposure effects. All functions can be applied separately as well as combined. The package "Matching" was developed by Sekhon (2011) for "Multivariate and Propensity Score Matching" (https://cran.r-project.org/web/packages/Matching/index.html). This package performs multivariate and PS matching and finds optimal balance based on a genetic search algorithm. The package "MatchIt" was developed by Kosuke Imai (Ho et al. 2007) for parametric causal inference (https://gking.harvard.edu/matchit). This package selects matched samples of the original treated and control groups with similar covariate distributions. It can be used to match exactly on covariates, match on PSs, or perform a variety of other matching procedures.

In this chapter, we make use of the functionalities of R package "nonrandom" to illustrate the PS analysis with the German Breast Cancer data to evaluate the medication effect of TMX on tumor size while adjusting for confounding covariates of "age", "meno", and "nodes".

As the first step, we load the R package and the data into R using following R code:

```
# Load the R library
library(nonrandom)
# load the data
newbreast = read.csv("newbreast.csv", header=TRUE)
```

> Estimate the PS: We estimate the PS through the logistic regression model, which is implemented by R function "pscore" in this "nonrandom" package.

```
> # Estimate ps
> newbreast.ps = pscore(data = newbreast,
           formula = TMX~age+meno+nodes)
> # Print the summary of PS
```

```
> summary(newbreast.ps)
Call:
glm(formula=formula,family = family, data = data, na.action = na.action)
```

Deviance Residuals:

Min	1Q	Median	3Q	Max
-1.4077	-1.0483	-0.6715	1.1952	1.9343

Coefficients:

	Estimate	Std. Error	z value	Pr(>|z|)
(Intercept)	1.792207	0.788185	-2.274	0.02298 *
age	0.027881	0.013054	2.136	0.03269 *
menoPreMeno	-0.821168	0.266406	-3.082	0.00205 **
nodes	0.002946	0.015100	0.195	0.84530

```
Null deviance: 895.38 on 685 degrees of freedom
Residual deviance: 836.01 on 682 degrees of freedom
AIC: 844.01
Number of Fisher Scoring iterations: 4
```

As seen from the output, the "pscore" function is in fact to call "glm" for logistic regression with "TMX" as the outcome variable and "age", "meno", and "nodes" as the independent variables. With the estimated PS, we can examine the PS distributions for each treatment group (i.e. with and without TMX) using the R function "plot.pscore" as follows:

```
># plot the distribution of PS
>plot.pscore(newbreast.ps, main = "German Breast Cancer Study",
         with.legend = TRUE,par.1 = list(lty=1, lwd=2),
    par.0 = list(lty=3, lwd=2), xlab = "PS", xlim=c(0,0.8))
```

PS Balance: Based on the estimated PS, we balance the covariate imbalance distributions between treatment groups embedded in the observational study. Stratification and matching by PS are the two most commonly used methods to eliminate these imbalances to mimic data situation as in randomized control trials (RCTs). In PS stratification, patients are grouped in the same strata with similar or equal values for their PSs.

The R function "ps.makestrata" is used for this purpose with options "stratified.by" or "breaks" to define the stratification variable in the data or to stratify by the specified breaks. We illustrate two ways to specify the "breaks" as follows. The first is simply to specify 3 strata with "breaks=3" and the second is to use "quantile" function in R to specify 5 strata with equal quantiles of 20%:

```
># The first: Stratification with 3 Strata
> newbreast.str.b3=ps.makestrata(object=newbreast.ps,
breaks=3)
> #Print the 3-strata stratification
> newbreast.str.b3
    Stratified by:  pscore
    Strata information:
       Strata bounds     n     n  (proportion)
    1  [0.117,0.288]  290        0.423
    2  (0.288,0.458]  149        0.217
    3  (0.458,0.629]  247     0.36
># The second: Stratification with 20% quantile
> newbreast.str.b5=ps.makestrata(object=newbreast.ps,
        breaks=quantile(newbreast.ps$pscore,seq(0,1,0.2)),
                    name.stratum.index = "stratum")
># Print the 5-strata Stratification
> newbreast.str.b5
    Stratified by:  pscore
    Strata information:
       Strata bounds     n     n  (proportion)
    1  [0.117,0.205]  138        0.201
    2  (0.205,0.241]  138        0.201
    3    (0.241,0.45]  136        0.198
    4    (0.45,0.493]  138        0.201
    5  (0.493,0.629]  136        0.198
```

In PS matching, one or more untreated patients are matched to the treated patients (or vice versa) based on their estimated PS to balance their covariate imbalances. In this situation, the matched patients have a similar or an identical estimated PS. This PS matching is accomplished by R function "ps.match" with matching "ratio" and a pre-specified maximum distance, defined as "caliper", between the estimated PS from treated and untreated patients as follows:

```
> # PS matching
> newbreast.m = ps.match(object = newbreast.ps,
                ratio = 1, caliper=0.5, setseed = 333)
> # Print the summary of PS matching
> summary(newbreast.m)
Matched by:  pscore
Matching parameter:
     Caliper size:   0.5
     Ratio:    1.0
     Who is treated?  1.0
Matching information:
     Untreated to treated? TRUE
     Best match?    TRUE
Matching data:
     Number of treated obs.:                      246
     Number of matched treated obs.:              246
     Number of untreated obs.:    4                40
     Number of matched untreated obs.:            246
     Number of total matched obs.:                492
     Number of not matched obs.:                  194
     Number of matching sets:                     246
     Number of incomplete matching sets:            0
```

After the PS stratification and matching to balance the covariate imbalances between the treatment groups, verification of the covariate balance between treatment groups after stratification or matching by PS is conducted. This verification process is usually done by graphical illustration followed by the statistical tests.

The R function "dist.plot" is implemented for graphical balance check (i.e. to check the covariate distributions by treatment groups). For the German breast cancer data, we can call this function to check covariate distributions of the three covariates "meno", "age" and "nodes", using the following code chunk:

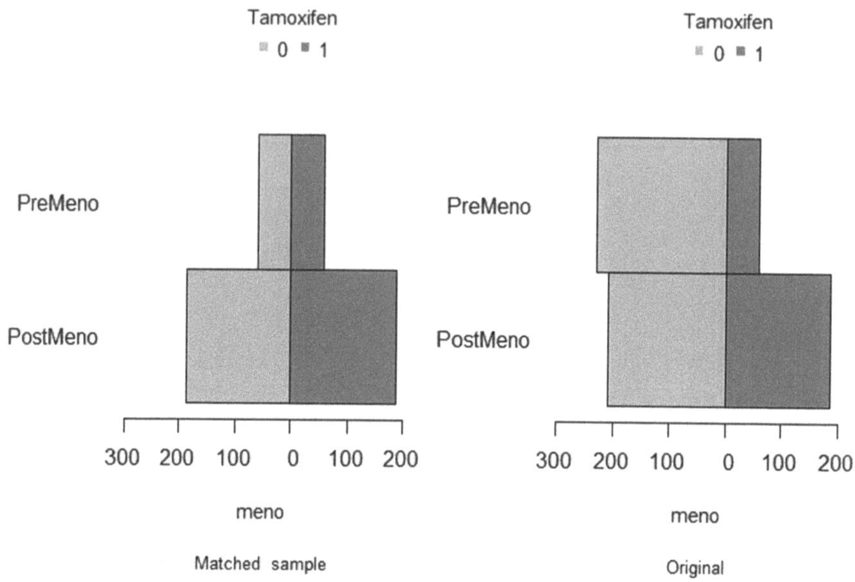

FIGURE 23.2
PS Distribution by Treatment Group before and after Matching for Covariate "meno".

```
># For Covariate-Meno
>newbreast.meno =dist.plot(object = newbreast.m,
                  sel = c("meno"),plot.type=2,
                  compare= TRUE,legend.title = "Tamoxifen")
```

```
># For Covariate=age
> newbreast.age = dist.plot(object = newbreast.m,
                  sel = c("age"), plot.type = 2,
                  compare = TRUE, legend.title = "Tamoxifen")
```

```
># For Covariate: nodes
>newbreast.nodes = dist.plot(object= newbreast.m,
                  sel= c("nodes"), plot.type = 2,
                  compare = TRUE, legend.title = "Tamoxifen")
```

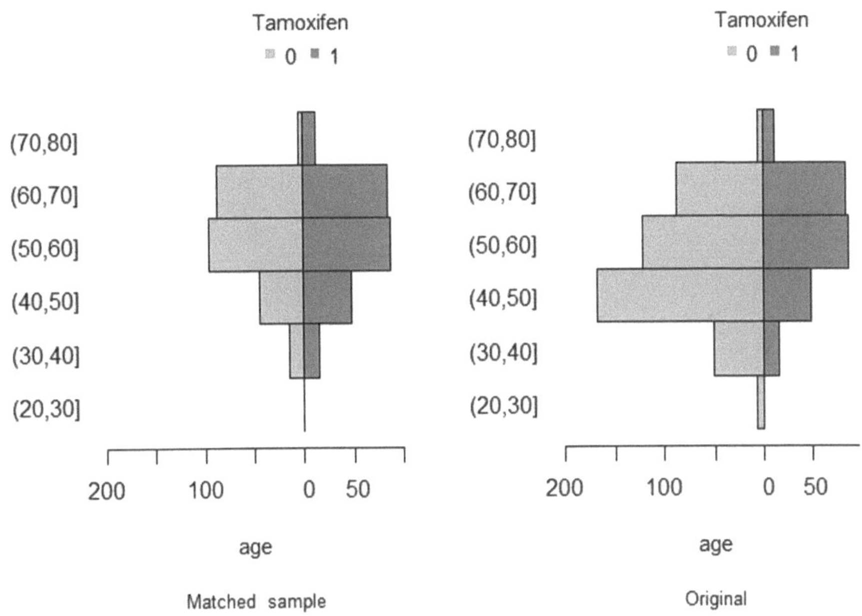

FIGURE 23.3
PS Distribution by Treatment Group before and after Matching for Covariate "age".

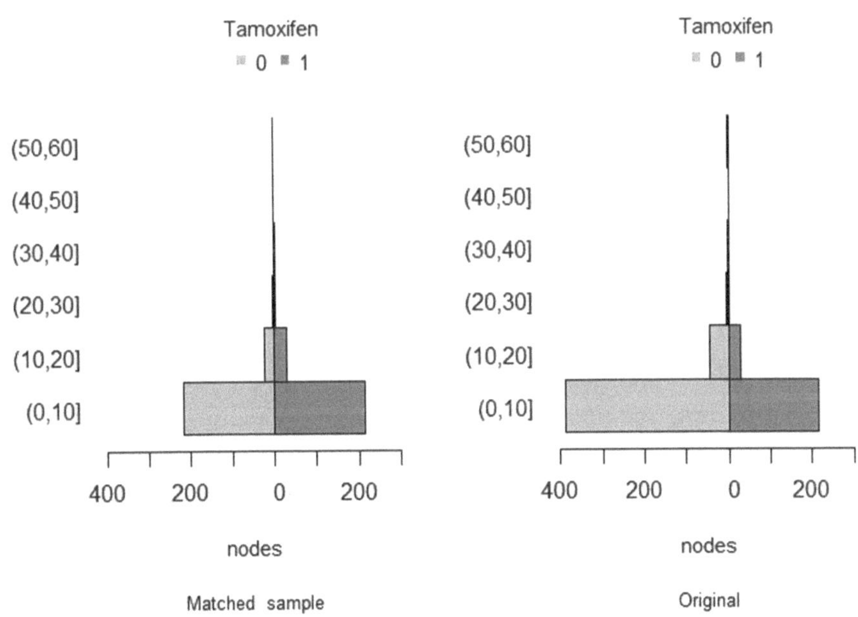

FIGURE 23.4
PS Distribution by Treatment Group before and after Matching for Covariate "nodes".

Figures 20.2–20.4 show that the PS are fairly balanced by the three covariates. As seen from Figure 20.2, the PS are balanced very well for "meno", and there are some differences for "age" and "nodes". We can further perform a statistical test to test whether the differences potentially visible in the graphics are statistically significant. The R function "ps.balance" can be called for this test, which can be illustrated as follows:

```
> # Test for balance using standardized differences
> newbreast.m.bal = ps.balance(object  = newbreast.m,
                               sel = c("age","meno","nodes"),
                               method  = "classical")
> # Print the summary of the test
> newbreast.m.bal
Summary of balance check:
                            Before: no bal (0) Before: bal (1)
    After: no bal (0)                       0               0
    After: bal (1)                          2               1
Detailed balance check (overall):
            age meno nodes
    Before   0    0    1
    After    1    1    1
Detailed balance check:
[p-values from tests (significance level: 0.05)]
               age     meno     nodes
    Before       0        0       0.665
    After    0.736        1       0.589
    Test         t    chi^2        t
```

From this test, the PS are balanced after the matching for all three covariates, even though "nodes" was balanced before the matching.

Estimation of treatment effect based on PS: The ultimate goal in PS analysis is to estimate the treatment effect with the PS. For the German breast cancer data, we want to estimate the medication (i.e., TMX) effect based on the PS. The R function "ps.estimate" is implemented for this purpose. We use the PS matched object "newbreast.m" for this illustration.

The R code is as follows:

```
># Estimate medication effect
> newbreast.est <- ps.estimate(object  = newbreast.m,
                              resp    =    "size",
                              adj   = c("age","meno","nodes"),
                              regr      = size~TMX+age+meno+nodes)
```

In the R code chunk of "ps.estimate", the "object" is the PS matched R object "newbreast.m" with the response variable "size". The option adj = c("age","meno","nodes") is used for additional adjustment for covariates in matched data. The default is NULL (i.e., no additional adjustment for covariates is needed). The option regr=size~TMX+age+meno+nodes is to report the classical regression estimation.

```
># Print the summary
> summary(newbreast.est)
Summary for effect estimation
Treatment/exposure: TMX
Outcome: size
Effect measure: effects
Effect estimates:
             effect    SE[effect]   [95%-CI[effect]]
Crude        -0.818    1.1385       [-3.049, 1.413]
Matching
Unadjusted   -0.797    1.1667       [-3.084, 1.49]
Adjusted     -0.977    1.0921       [-3.117, 1.163]
Regression   -0.548    1.124        [-2.751, 1.655]
```

In the output, the "Crude" is the crude effect of TMX from the classical t-test without any covariate adjustment where the effect is estimated at -0.818 with SE of 1.1385 and 95% CI of [-3.049,1.413]. With the PS matching (i.e., Matching), the "Unadjusted" (i.e., without addition adjustment of the three covariates) treatment effect is -0.797 with SE of 1.1667 and 95% CI of [-3.084,1.49]. Since we specified adj = c("age","m eno","nodes") in the "ps.estimate", the "Adjusted" treatment effect is -0.977 with Se of 1.0921 and 95% CI of [-3.117,1.163]. The "Regression" is to give the traditional analysis of covariance with the three covariates, which gives the treatment effect of -0.548 with SE of 1.124 and 95% CI of [-2.751, 1.655]. Results from all these analyses indicate that the medication effect (i.e. TMX effect) on breast cancer tumor size is not statistically significant.

23.5 Discussions

Selection bias is not uncommon in observational studies. The PS makes an adjustment for selection bias. When used with traditional risk adjustment techniques, it reduces bias and provides a better description of the effect of covariates on outcomes. Several studies have demonstrated that PS matching eliminates a greater proportion of the systematic differences in baseline characteristics between treatment and control than stratification on the PS or covariate adjustment using the PS (Austin 2009; Austin et al. 2007; Austin and Mamdani 2006).

23.6 Exercises

Use the German Breast Cancer data to answer the following questions. Does hormone affect grade in breast cancer?

Treatment: Hormone [tamoxifen or no tamoxifen]

Outcome: Grade

Confounders: age, menopause, and tumor size

References

Austin, P. C. (2009). Type I error rates, coverage of confidence intervals, and variance estimation in propensity-score matched analyses. *International Journal of Biostatistics*, *5* (1), 13. https://doi.org/10.2202/1557-4679.1146

Austin, P. C. (2011). An introduction to propensity score methods for reducing the effects of confounding in observational studies. *Multivariate Behavioral Research, 46*, 399–424.

Austin, P.C., Grootendorst, P., and Anderson, G. M. (2007). A comparison of the ability of different propensity score models to balance measured variables between treated and untreated subjects: A Monte Carlo study. *Statistics in Medicine, 26*, 734–753. https://doi.org/10.1002/sim.2580

Austin, P. C. and Mamdani, M. M. (2006). A comparison of propensity score methods: A case-study estimating the effectiveness of post-AMI statin use. *Statistics in Medicine, 25*(12), 2084–2106. https://doi.org/10.1002/sim.2328

Federal Drug Administration (FDA) (2020). Center for Drug Evaluation and Research: Tamoxifen Information: Reducing the incidence of breast cancer in women at high risk. https://web.archive.org/web/20070619012859/ www.fda.gov/cder/news/tamoxifen/ (2005). Accessed 26 July 2020.

Guo, S. and Fraser, M. W. (2015). *Propensity Score Analysis: Statistical Methods and Applications*. Washington, D.C: Sage.

He, H., Wu, P., and Chen, D. G. (2016). *Statistical Causal Inferences and their Applications in Public Health Research*. New York: Springer.

Ho, D., Imai, K., King, G. and Stuart, E. (2007). Matching as nonparametric preprocessing for reducing model dependence in parametric causal inference. *Political Analysis, 15*, 199–236.

Pan, W. and Bai, H. (2015). *Propensity Score Analysis: Fundamentals and Developments*. New York: Guilford Publications.

Royston, P. and Sauerbrei, W. (2008). *Multivariable Model-Building: A Pragmatic Approach to Regression Analysis Based on Fractional Polynomials for Modelling Continuous Variables*. Chichester: Wiley.

Rubin, D. B. (1974). Estimating causal effects of treatments in randomized and nonrandomized studies. *Journal of Educational Psychology, 66*, 688–701.

Rubin, D. B. (1976). Inference and missing data (with discussion). *Biometrika, 63*, 581–592.

Rubin, D. B. (2004). Direct and indirect causal effects via potential outcomes. *Scandinavian Journal of Statistics, 31*, 161–170.

Rubin, D. B. (2005). Causal inference using potential outcomes: Design, modeling, decisions. *Journal of the American Statistical Association, 100*, 322–331.

Sekhon, J. S. (2011). Multivariate and propensity score matching: Software with automated balance optimization: The matching package for R. *Journal of Statistical Software, 42*(7), 1–52.

Index

Absolute Index, 172
Adjusted GFI (AGFI), 172
Adjusted hazard risk, 57
Akaike information criterion (AIC), 183–184, 212
Associated equal weighting (cW), 27
Associated statistical inference, 61, 153
Asthma Data, 158–160
Asthma respiratory function (RF), in adolescents, 158
Asymptotic variance, computation of, 15–16

Baseline Model Chi-Square estimate, 172
Bayesian data analysis
  binary outcome
    by using R packages, 78–79
    by using SAS programming for binary response, 74–78
  continuous outcome
    by using R packages, 67–74
    by using SAS program, 64–67
Bayesian information criterion (BIC), 183–184
Bayesian modeling, 60–63, 89
  advantages and disadvantages of, 80–81
  assessing Markov-Chain convergence, 63
  Bayesian updating and inference, 61–62
  Markov-Chain Monte-Carlo (MCMC) methods, 62–63
  obtaining information from posterior, 62–63
  principle of, 61
  visual analysis via trace plot, 63
Bayesian paradigm
  overview of, 59
  posterior probability in, 60
  public health NHANES data, 59
Bayesian posterior and predictive inference, 156
Bayes logistic regression, 75, 79
Bayes theorem, 60–61
  application of, 61
Bayes, Thomas, 60
Bentler-Bonett Normed Fit Index (NFI), 173
Bernoulli distribution, 16
Binomial distribution, 75, 123
Bivand, Roger, 156
Breast cancer data
  German Breast Cancer data, 224–225
  with measure patterns of change in personality assessment, 181–182
Breast cancer, prevention of, 224
Burn-in period, determination of, 63, 66, 77

Child–parent psychotherapy, 209
Child welfare system, 207
Chinese Health Questionnaire, 12-item (CHQ), 181
Chinese Social Adjustment Scale, 33-item, 182
Chi-square distribution, 113
Chi-Square Test of Model Fit, 167, 175, 177
Chlorine data, 97–98
Christensen, Ole F., 156
Classical regression, 154–155
  coefficient of determination in, 114
Classical statistical modeling, 134
Comparative Fit Index (CFI), 167, 177, 194, 196, 199
Computer information systems, 39
Conditional probability, 60
  concept of, 60
Confirmatory factor analysis (CFA), 162–163
Convergence diagnostics, 62–63
Correlated binary responses, joint modeling for, 7–13
Covariance structure analysis, 171, 173
Cox proportional hazard model, 49, 51, 57, 91–92
Cox regression model, 48, 214

CRAN Task View, 156
Credible intervals, 62, 77
Credible sets, 77
Cumulative distribution function (cdf), 44
Cumulative hazard function, 45

"DATA" command, 165
Decision-making, 136
  public health, 160
Degrees of freedom, 90, 113, 185
DerSimonian–Laird's method of moments (MM), 112–113, 126
Devereux Early Childhood Assessment for Infants and Toddlers (DECA-I/T), 209
Diggle, Peter J., 156
Disability resources and services (DRS), 38
Dist.plot (R function), 233
"Drug" effect, 90–91, 94

Econometrics, 3, 45
Effective sample size (ESS), 67, 78
Empirical variogram, 152–153
Evidence-based parenting interventions, 209
Extrapolated volume (NEV), 160, 162

Fitted spherical variogram model, 152
Fixed-effects MA (RE-MA) model, 108, 110–111, 116, 126
Food and Drug Administration (U.S.), 224
Forced vital capacity (FVC), 160, 162
Fox, John, 174
Frequentist hypothesis testing, 60
Frequentist paradigm, 59, 61

Gaussian random field, 156
Generalized estimating equations (GEEs), 2
  for multiple outcomes, 14–15
Generalized linear mixed-effects model (GLIMMIX), 30
Generalized method of moments (GMM) model, 1
  applications and extensions of, 7, 22
  estimation for time-dependent covariates, 1–3, 13–16
    advantages and limitations of, 5
    advantages of the joint models in, 22
    asymptotic variance for, 15–16
    classification of, 4–5
    efficacy in, 4
    framework of, 4
    hypothesis testing for moment conditions, 5
    marginal models, 2
    for multiple outcomes, 14–15
    multivariate longitudinal models, 3
    simultaneous approach, 3
    single-parameter models for, 4
  identification of valid moment conditions, 6–7
  joint modeling for correlated binary responses, 7–13
  numerical example of, 17–21
    modeling of smoking and social alcohol use, 17–18
    modeling of smoking, social alcohol use, and obesity, 18–21
  partitioned model framework, 6
GeoR (R package), 156
GeoRglm (R package), 156
Geostatistical regression model, 153–155
German Breast Cancer data, 224
  propensity score analysis of, 224–225
Geweke's diagnostic, 77
Gibbs sampler, 62
GitHub repository, 16
Goodness of Fit Index (GFI), 172
Goodness-of-fit (GOF) statistic, 160, 167, 175
Grade point averages (GPAs), 38
Growth curve modeling, 181, 183, 191, 205–207, 209, 211
Growth mixture modeling, 206–209, 214, 220
Gstat (R package), 156

Hazard function, 45, 47, 49, 87, 214
Hazard rate, 45, 56, 91, 213
Hazard ratio, 48–50, 54, 56–57, 88
Healthcare system, 25

Health-related behaviors, longitudinal study of, 17
Higher-order interactions, 22
Highest probability density (HPD), 62
H-index, 113
Hospital doctor patient (HDP) data
  with multiple membership, 26–28

I^2 index, 113–114
Illinois Department of Children and Family Services (IDCFS), 208
Imai, Kosuke, 230
Incremental Index statistics test, 173
Integrative data harmonization (IDH) framework, 206

JMbayes (R package), 88–89
Joint growth mixture modeling (JGMM), 209, 214–215

Kaplan–Meier estimator, 46–47, 51
  survival curve using SAS procedure, 51
  time-to-death study with SAS programming, 51–53
Kaplan–Meier Graph, 47

Latent class analysis, 205, 207
Latent class growth curve (LGC) modeling
  for "mood" and "social adjustment", 191–197
    with covariates, 197–202
Latent class growth models (LGMM), 191, 202, 207
Latent class trajectories, 215–216, 220
  profile analysis of, 217–219
Latent constructs, validity of, 162–163
Latent growth mixture joint modeling (LGMJM), 205
  application in real-world social intervention settings, 208
  benefits of, 220
  components of, 207–208
  demonstration using R for data analyses
    analysis framework, 209–210
    longitudinal data, 210–213
    time-to-event data, 213
  joint modeling of
    longitudinal and time-to-event data, 214–215
  latent-class trajectories identified by, 216
  motivation for, 206–208
    heterogeneous trajectories among unobserved subpopulations, 206–207
    outcome-dependent missingness, 207
  time-to-event data, 215–217
    demonstration using R, 213
    joint modeling of, 214–215
Lavaan (R package), 158, 174
Lcmm (R package), 214
Least-squares estimation, in the linear regression, 153
Likelihood ratio chi-square test, 90, 211, 212, 215
Likert scale, 4-point, 181
Linear growth curve modeling, 183
Linear mixed effects (LMEs) model, 34, 85, 92
  random-intercept, 90
  R codes for, 89, 211
Linear predictors, 48
Linear regression coefficients, of factor loadings, 159
Linear regression model, 97
  least-squares estimation in, 153
  without spatial autocorrelation, 147–149
Log hazard, 46, 50, 91, 213
Logistic regression model, 21, 49, 59, 124, 224
Longitudinal and survival data analysis
  HIV/AIDS data and preliminary data analysis, 84–85
  results of, 89–93
    CD4 longitudinal data, 89–91
    joint modeling of longitudinal CD4 and time-to-death, 92–93
    separate Cox proportional hazards regression, 91–92
  separate modeling of
    longitudinal continuous data, 86–87
    time-to-event data, 87–88
  trends for CD4 measures, 86

Longitudinal continuous data, 86, 93
joint modeling with time-to-event data, 88–89
Longitudinal data analysis, 205
Longitudinal linear mixed-effects models, model selection statistics for, 90
LU decomposition technique, 156

Marginal models, for single responses with time-independent covariates, 2
Markov-Chain Monte-Carlo (MCMC) methods, 62–63, 88, 156
graphical representation of, 92
illustration for the parameters in the longitudinal CD4 component, 92
"Markov-Chain" simulation, 62, 78
MatchIt (R package), 230
Materm variogram, 152
Maximum likelihood estimation (MLE), 112
analysis of, 226
Maximum likelihood method, 46
analysis of, 54
Maximum likelihood ratio test, 184–185, 188
Meta-analysis (MA), 108
fixed-effects meta-analysis models, 110–111
metadata on Bacillus Calmette-Guerin Vaccine Clinical Trials, 109
random-effects MA (RE-MA) model, 108, 111–113
testing and quantifying the between-study heterogeneity, 113–114
using R package
calculation of effect size, 124–125
data and R package "metafor", 122–123
defining of effect size, 123–124
fixed-effects meta-analysis, 126
meta-regression, 128–131
random-effects meta-analysis, 126–128
using SAS program
fixed-effects meta-analysis models, 116–117
meta-data and effect size, 115–116
meta-regression, 120–122
random-effects meta-analysis models, 117–120
Metafor (R library), 109, 122–123
Meta-regression, 114–115, 120–122, 128–131
forest plot after, 131
Metropolis–Hastings (MH) algorithm, 62
Meuse River dataset, 135–136, 137
geostatistical regression model for analysis of, 153–155
spatial analysis using R package
estimation of spatial autocorrelation in the residuals, 150–153
graphical analysis, 146–147
linear regression without spatial autocorrelation, 147–149
Missing not at random (MNAR), 85, 208
Mixed-effects modeling, 93, 210
"Model" command, 166, 174
Moderators, 114
Moment conditions, hypothesis testing for, 5
"Monte-Carlo" simulation, 62
Mood
data analysis of, 184–187
longitudinal linear growth curve models with, 184
intercept of, 196
joint LGC modeling for, 191–197
with covariates, 197–202
measurement of, 181
MPus Program, 164–169
Multiple membership data
analysis using R package
data format, 34
loading the R Library and the data, 34
random intercept LME model, 35–37
analysis using SAS
data format, 29–30
procedure of, 30–33
exercises of, 38–39
hospital doctor patient (HDP) data, 26–28
two-level linear mixed-effects models for, 28–29

Multiplication of probabilities, law of, 47
Multivariate longitudinal data, 3

N-acetyl-S-(benzyl)-L-cysteine, 160
National Comorbidity Study, 17
National Health and Nutritional Examination Surveys (NHANES), 160
National Longitudinal Study of Adolescent to Adult Health (Add Health), 2, 17
Newton–Raphson method, 10
New York Leukemia dataset, 135–136
  geostatistical regression model for analysis of, 153–155
  spatial analysis of
    using R package, 146–153
    using SAS program, 140–146
NHANES data, analysis of, 164–177
  using MPus program, 164
    SEM program for chapter 20-CFA analysis, 165–168
    using the fitted CFA model for full SEM, 168–169
  using R program, 174–177
  using SAS program, 169–173
Nlme (R package), 89
Nonlinear optimization, 10
Nonlinear regression models
  chlorine data, 97–98
  complexity associated with, 98–99
  conditions for fitting, 99–101
  normal probability plot, 100
  prediction, 101
  research interest, 97
  residual *versus* predicted plot, 100–101
Nonparametric model, for estimating the survival function, 46–47
Nonrandom (R package), 230
Normal distribution, 16, 29, 37, 62
Null hypothesis, 5, 60, 113
Nurturing Parenting Program, 209

Obesity, predictors for, 21
Odds ratio (OR), 124
Online documentation, 134
Ordinary least-square (OLS) regression, 183
  with spatially autocorrelated errors, 142–144
  without spatial autocorrelation and weighting, 140–141
Outcome-dependent missingness, 86, 94, 206–208, 220
"OUTPUT" command, 167

Parametric causal inference, 230
Parametric linear regression model, 46
Parsimony Index, 172
Partial-likelihood estimation, 50
Path analysis, 158, 164, 174
Peak expiratory flow (PEF), 160, 162
Posterior distribution of parameter, 61–62
Posterior probability, 60–62
Probability distribution, 60, 62
Probability of surviving, 43–44
PROC CALIS procedure, 158, 169, 172–173
Proc GENMOD, 74
Propensity score (PS) analysis, 223, 224–225
  estimation of treatment effect based on, 228, 236
  R program for, 230–238
  SAS program for, 225–229
  stratification of, 233
Proportional hazard model, 48, 50, 57
Ps.makestrata (R function), 232

Q-statistic, 113, 126

Random-effects MA (RE-MA) model, 108, 111–113, 117–120, 127, 128
Random intercept LME model, 35–37
Randomized control trials (RCTs), 227, 231
Real data
  data source, 208–209
  outcome variables, 209
Regression coefficients, 77
  interpretation of, 49
Restricted maximum likelihood (REML) method, 112
R function "fit.variogram", 152
Ribeiro, Paulo J., Jr, 156

Risk difference (RD), 123
Root mean square error of
approximation (RMSEA), 168, 172, 177, 196
Rosseel, Yves, 174
R statistical programming language, 87, 211

Sem (R package), 174
Semiparametric maximum likelihood
estimation, 88
Semivariogram, in spatial statistics, 151
Separate Cox proportional hazards
regression model, 91
Separate linear mixed-effects model, 90
Sequential clinical trials, 60
Sidik–Jonkman (SJ) estimator, 112
Simpler distribution, 62
Simulation index, 63
Social adjustment (SOCAD)
data analysis for, 187–191
intercept of, 196
joint LGC modeling for, 191–197
with covariates, 197–202
longitudinal data analysis for
analytical approach to, 183–184
measurement of, 182
Social alcohol use, model for
prediction of, 20
Social sciences, 45
Socioemotional well-being, 209
Software development, 28, 156
Spatial autocorrelation, 134, 139–141, 143–145, 150, 152–153
Spatial dependence, 134, 153, 155
Spatial statistical analysis
goal of, 139
of Leukemia Data using SAS, 140–146
OLS without spatial autocorrelation
and weighting, 140–141
OLS with spatially autocorrelated
errors, 142–144
weighted linear regression without
spatial autocorrelation, 141–142
weighted LS with spatially
autocorrelated errors, 144–146
meaning of, 139
of Meuse River dataset using
R package
estimation of spatial autocorrelation
in the residuals, 150–153
graphical analysis, 146–147
linear regression without spatial
autocorrelation, 147–149
models in, 136–139
Spatial statistics, 134
Spatial survey sampling, 134
Standardized root mean square residual
(SRMR), 168, 177, 196
Statistical joint modeling, 85–89
Statistical modeling
paradigms for statistical estimation
and inference
Bayesian paradigm, 59
frequentist paradigm, 59
Structural equation modeling (SEM), 160, 191
confirmatory factor analysis, 162–163
goodness-of-fit criteria, 160
versus regression analysis, 163–164
theoretical constructs of, 158
Student residency classification, 38
Supply chain management, 39
Survival data, statistical modeling of
based on a response variable "time", 44
censored *versus* uncensored
observations, 43
functions of
hazard function, 45
survival function, 44–45
HIV/AIDS data, 42
principles of, 42–44
research interest/question, 41–42
right censoring, 43–44
Survival function, 44–45
estimates of, 45
illustration of, 45
Survival models, 45–51
assumptions, 49–50
Cox regression diagnostics, 50
Cox's proportional hazards regression
semiparametric model, 47–49
data analysis, 51–54
estimation, 50
extended Cox proportional hazard
model, 50–51
interpretation of regression
coefficient, 49

Kaplan–Meier estimator, 46–47
parametric and nonparametric models of, 45
parametric model, 46
use in analysis of medical and biological sciences, 45

Tamoxifen (TMX), 224
Three Title IVE Waiver (IB3) demonstration project, 208
Tidyverse (R library), 184
Time, analysis of, 182
Time-dependent covariates
advantages and limitations of, 5
classification of, 4–5
estimation for, 1–3
GMM framework for, 4, 13–16
marginal models, 2
multivariate longitudinal models, 3
simultaneous approach, 3
extended classification of systematic component models for, 5–7
identification of valid moment conditions, 6–7
model applicability and extensions, 7
partitioned GMM model framework, 6
extended Cox model with, 50
hypothesis testing for moment conditions, 5
regression model, 7
time-varying effects for, 5
Time-invariant covariates, 94, 184
Time-invariant predictors, 182, 183, 185
Time-to-death study, 51–54, 91
Cox proportional hazard regression, 53–54
for HIV/AIDS response for patients, 89
Kaplan–Meier method, 51–53
LIFETEST procedure, 52
number of event and censored values, 53
PHREG procedure, 53
results of joint modeling of longitudinal CD4, 92–93
with R programming, 54–56
with SAS programming, 51–54
Time-to-event data, 94, 207–208
joint modeling with longitudinal continuous data, 88–89
latent growth mixture joint modeling (LGMJM) of, 215–217
separate modeling of, 213
Time-to-permanence dataset, 208
"Tobacco" effect, 162
Trace plots, 63
Trichloroethylene (TCE), 135
Tucker-Lewis fit index (TLI), 168, 177, 196

"VARIABLE" command, 165
Variogram, 150–151
residual, 154
Variogram cloud, from the residuals, 151
Volatile organic compounds (VOCs), 158–159

X predicts Y (predictor–outcome framework), 163

Z-statistics, 5, 154

FSC
www.fsc.org
MIX
Papier | Fördert
gute Waldnutzung
FSC® C083411

Kolibri 360 GmbH
Ferdinand-Jühlke-Straße 7
99095 Erfurt, Deutschland
produktsicherheit@kolibri360.de